Marine Shell Trade and Classic Teotihuacan, Mexico

Charles C. Kolb

BAR International Series 364
1987

B.A.R.

5, Centremead, Osney Mead, Oxford OX2 0DQ, England.

GENERAL EDITORS

A.R. Hands, B.Sc., M.A., D.Phil.
D.R. Walker, M.A.

BAR-S364, 1987 'Marine Shell Trade and Classic Teotihuacan, Mexico'

© Charles C. Kolb, 1987

The author's moral rights under the 1988 UK Copyright,
Designs and Patents Act are hereby expressly asserted.

All rights reserved. No part of this work may be copied, reproduced, stored, sold, distributed, scanned, saved in any form of digital format or transmitted in any form digitally, without the written permission of the Publisher.

ISBN 9780860544722 paperback
ISBN 9781407346397 e-book
DOI https://doi.org/10.30861/9780860544722
A catalogue record for this book is available from the British Library
This book is available at www.barpublishing.com

TABLE OF CONTENTS

Page

iii	Table of Contents
vi	List of Figures
ix	List of Appendices
x	Preface
1	Chapter One: Classic Period Teotihuacan
1	Introduction to the Research
5	Teotihuacan: City, "State," and "Empire"
5	The City and Its Craft Specializations
10	Merchants: "Foreign" and Domestic
12	Teotihuacan Obsidian and Other Resources
13	The Urban Social System
15	The Rural and Provincial Regions
16	The "State" and "Empire"
19	Chapter Two: Major Mollusks
19	Classifications and Marine Faunal Provinces
20	*Spondylidae*
23	*Strombidae*
25	*Fasciolariidae*
27	Chapter Three: Marine Mollusks in the Art and Artifacts at Teotihuacan
27	Marine Representations in Mural Art
33	Marine Representations in Lithic Sculptures
39	Marine Mollusks and Ceramic Artifacts
45	Chapter Four: Mollusk Specimens at Teotihuacan
45	Burials and Caches at Urban Residences
45	Linne's Excavations
47	Tlamimilolpa Burial 1
47	Tlamimilolpa Burial 2
49	Tlamimilolpa Burial 4
49	Tlamimilolpa Burial 5
49	Tlamimilolpa Burial 13
49	Tlamimilolpa Cache 1
50	Tlajinga 33 Excavations
52	Tlajinga 33 Burial 6
52	Tlajinga 33 Burial 10
52	Tlajinga 33 Burial 14
52	Tlajinga 33 Burial 15
53	Tlajinga 33 Burial 17a
53	Tlajinga 33 Burial 25
53	Tlajinga 33 Burial 44

53	Tlajinga 33 Burials 50a through 50e
53	Tlajinga 33 Burial 56
54	Tlajinga 33 Burial 57
54	Tlajinga 33 Feature 11
54	Tlajinga 33 Feature 52
55	Noguera and Sejourne's Excavations
56	Other Mollusks at Teotihuacan
57	Chapter Five: Santa Maria Maquixco el Bajo: TC-8
57	Introduction
60	The Prehistory of Maquixco el Bajo
63	The Colonial and Recent History of Maquixco el Bajo
65	Xolalpan Phase TC-8
65	The Temple Pyramid
65	Structure TC-8:1-2
67	Structure TC-8:3
68	Structure TC-8:4
69	Summary and Interpretation of Xolalpan Phase TC-8
78	Chapter Six: The Shells of Maquixco and Other "Rural" Sites
78	Introduction
78	TC-8:3 Platform Altar _Ofrendas_
81	TC-8:3 Room 2 _Bodega_
82	TC-8:3 "Foreign" Connections
83	Preliminary Hypothesis
84	A Summary of Teotihuacan Valley Rural Site Mollusks
90	Chapter Seven: Major Mollusk Distributions
90	Introduction
90	_Spondylidae_
93	_Strombidae_
95	_Fasciolariidae_
99	Chapter Eight: Mollusks and Their Uses
99	Nutrition
99	Utilitarian and Construction
100	Shellfish Dyes
101	Shell Artifacts
102	Shell Trumpets
105	Chapter Nine: The Shell Procurement Network
105	Introduction
106	Some Exchange System Parameters
107	Professional Merchants
109	"Ports of Trade"
111	Potential Teotihuacan Commercial Routes to the Coasts
111	Gulf Coast Routes A - G
116	Pacific Coast Routes A - G

122	Chapter Ten: Classic Teotihuacan Commerce: The Role of Mollusks
122	Introduction
122	The Shell Procurement System - The Teotihuacan Connection
125	The Shell Procurement System - A Commercial Model
127	Appendices
179	Glossary of Architectural Terms
182	Glossary of Spanish and Nahuatl Terms
184	Bibliography
225	Personal Communications
226	About the Author

LIST OF FIGURES

Figure	Page	Title
1	2	The Mesoamerican Culture Area
2	3	The Basin of Mexico
3	4	Chronological Concordances for the Basin of Mexico
4	6	The Teotihuacan Valley During the Classic Period
5	7	Urban Teotihuacan During the Xolalpan Phase (ca 600 A.D.)
6	8	The Basins of Mexico, Tlaxcala and Puebla: Major Commercial Routes
7	21	Marine Faunal Provinces of the Mesoamerican Culture Area
8	28	Plan of the Palace and Patio of the Jaguars (76:N4W1)
9	30	Plan of the Atetelco Site (1:N2W3)
10	32	Plan of the Tetitla Site (1:N2W2)
11	34	Plan of the Tepantitla Site (1:N4E2)
12	35	Plan of the Yayahuala Site (1:N3W2)
13	36	Plan of the Zacuala Patios Site (2:N2W2)
14	37	Plan of the Zacuala Palacio Site (3:N2W2)
15	38	Plan of the Palacio de Quetzalpapalotl (2:N4W1)
16	40	Examples of Mold-made Ceramic _Adornos_
17	41	Example of a Composite Censer with _Adornos_ (Late Tlamimilolpa to Early Xolalpan Phases)
18	43	Example of a Granular Ware Ceramic Trumpet (Reconstructed) (Late Xolalpan Phase)

19	46	Plan of the Xolalpan Site (2A:N4E2)
20	48	Plan of the Tlamimilolpa Site (1:N4E4)
21	51	Plan of the Tlajinga 33 Site (33:S3W1)
22	58	Plan of the Archaeological Remains at the Santa Maria Maquixco el Bajo Site (TC-8)
23	59	Plan of the Santa Maria Maquixco el Bajo Site Excavations (TC-8:1-2, 3, 4)
24	73	Santa Maria Maquixco el Bajo Site Excavations (TC-8:1-2, 3, 4): Examples of *Spondylus calcifer* Carpenter, 1857 (Panamanian Marine Faunal Province)
25	74	Santa Maria Maquixco el Bajo Site Excavations (TC-8:1-2, 3, 4): Various Unidentified, Worked Marine Shell Specimens
26	75	Santa Maria Maquixco el Bajo Site Excavations (TC-8:1-2, 3, 4): Two Identified, Unworked Marine Shell Specimens (Panamanian Marine Faunal Province)
27	76	Santa Maria Maquixco el Bajo Site Excavations (TC-8:1-2, 3, 4): Three Identified, Unworked Marine Shell Specimens (Panamanian Marine Faunal Province)
28	77	Santa Maria Maquixco el Bajo Site Excavations (TC-8:1-2, 3, 4): *Unionidae* (Family), *Elliptio* (Genus) "Freshwater Clam" Worked Specimens
29	86	Venta de Carpio Site Excavation (TC-10:2): Three Unworked Marine Shell Specimens
30	87	Teotihuacan Valley Project Classic Period Surface Survey (TC-2, 8, 13, 40, 73, 91): Various Identified and Unidentified Marine Shell Specimens
31	92	*Spondilidae* Distributions in Mesoamerica
32	94	*Strombidae* Distributions in Mesoamerica

33	96	<u>Fasciolariidae</u> Distributions in Mesoamerica
34	112	Gulf Coast to Highlands Commercial Routes during the Classic Teotihuacan Period
35	117	Pacific Coast to Highlands Commercial Routes during the Classic Teotihuacan Period

LIST OF APPENDICES

Page

127	Index to the Appendices
128	I. Specific Mollusks Represented in Classic Period Teotihuacan Valley Rural Sites: Site, Genus and species
130	II. Summary of Specific Mollusks Represented in Classic Period Teotihuacan Valley/Sites: Classes and Marine Faunal Provinces Rural
131	III. Comments on Selected Mollusks Represented in Classic Period Teotihuacan Valley Rural Sites
133	IV. Mollusks Represented in Classic Period Basin of Mexico Sites
141	V. Marine Shell in Classic Period Sites in the Teotihuacan Valley: The Sites
142	TC-2
144	TC-8
154	TC-10
156	TC-13
159	TC-40
163	TC-46
168	TC-73
177	TC-91, TC-92, and TC-93

PREFACE

The Mesoamerican Culture Area was one of the six areas of the world where "civilizations" independently evolved. Four of these were in the Old World -- Mesopotamia or Southwestern Asia, the Indus River Valley or Asian Subcontinent, Northern China, and the Nile River Valley or Northeastern North Africa. In addition to the Mesoamerican area, the Central Andes or Peruvian Culture Area of South America also was a locus of the evolution of New World "civilization." The term "civilization" is used to mean an area which had the independent evolution of "cities" (civitas from the Greek) and an accompanying series of traits. The concept includes the presence of a high level of food production, the development of managerial skills to handle developing institutions (economic, social, and military), control of the means of production by "ruling" classes, professional merchant groups working within a highly developed exchange network, and full-time craft specialists freed from manual labor. "Civilizations" are also marked by dense populations, artistic styles elaborated by artists-craftspersons, monumental public buildings symbolic of the concentration of social surpluses, and tithe or tax in the form of surplus products made by craft producers and cultivators. Specialists were often responsible for the development of systems of recording and the exact or predictive sciences (arithmetic; geometry; astronomy; numerical notations; standards of time, space, and weight; calendrics; and writing systems). Craft specialization, exploitation of diverse ecological zones, commerce, resource concentrations, formal exchange networks (with tangible and intangible "goods," the latter including religious and political ideologies), and demographic growth are particularly salient criteria of the evolution of a "civilization" (Sanders and Price 1968).

Mesoamerica as a distinct Culture Area was initially defined by Paul Kirchhoff (1943), and was the locus of a number of high cultures or "civilizations." The Meseta Central (Central Plateau) saw the waxing and waning of the Classic Period Teotihuacan "state" and probable "empire" (ca 50-750 A.D.), followed by the Postclassic Toltec "state" (ca 800-1150 A.D.) and the Postclassic Aztec "state" and associated pan-Mesoamerican "Empire" (ca 1150 - 1520 A.D.). The Maya region, comprised of the Yucatecan Lowlands and Guatemalan Highlands, saw the evolution of the Classic (ca 300 - 600 A.D.) and later the Post Classic Maya (ca 900 - 1530 A.D.). The Classic Period Teotihuacan "state" and probable "empire" is the focus of this monograph.

Following the Second World War, archaeology emerged as a "modern" discipline with new theoretical orientations, elaborated field methodologies, and more analytical laboratory techniques. Paramount as "new" methods were chronometric dating, especially radiocarbon dating, and paradigms oriented to the ecological approach. In the field, ecologically oriented settlement pattern studies were developed. Archaeological data are, in the main, recovered through surface survey or reconnaissance and site excavations, and are supplemented by information derived through a detailed (extensive and intensive) analysis of the interrelationships between the natural and biological environments, the technical analysis of artifactual and biotic remains, the studies of human settlements, and the use of models derived from ethnohistoric and ethnographic studies in relation to physiographic and sociocultural factors. Site survey or surface reconnaissance can and does provide extensive diachronic and synchronic data through systematic inspection, as witnessed by the results of the surveys conducted in the Viru Valley, Peru, by Gordon R. Willey and the Institute of Andean Research. This pioneering settlement pattern study ultimately led to a conference in 1960 designed to coordinate settlement pattern and other anthropological projects in the Basin of Mexico, a key region within Mesoamerica.

As a result of the 1960 conference, specific and independent research projects were initiated by William T. Sanders and Rene F. Millon in the Teotihuacan Valley, a northeastern segment of the Basin of Mexico (Kolb 1979a:4, 50-55). Personnel of the "Teotihuacan Valley Project" (TVP) of The Pennsylvania State University, directed by Sanders, conducted archaeological surveys and test excavations in the rural areas of that Valley from 1960 - 1965, while the "Teotihuacan Mapping Project" (TMP) of The University of Rochester (Rochester, New York, U.S.A.), directed by Millon, simultaneously began an intensive survey with limited test excavations in the urban center, a task which continues to the present. Millon and his colleagues published a detailed series of maps locating all archaeological remains in the metropolis and "established" the urban boundaries (Millon 1973). The survey of the urban center necessitated a specialized mapping procedure to designate and locate individual "sites" (dwellings, apartment complexes, temples, etc.), hence, a grid system of 500 x 500 m^2 units was developed using the largest architectonic unit (the Ciudadela or Citadel) and the main north-south thoroughfare (the Miccaotli or so-called "Street of the Dead") as grid orientations. Individual "sites" within each grid were individually numbered; for example, the Tlamimilolpa urban residence complex excavated by Sigvald Linne (1942) received the designation of 1:N4E4, meaning site number "one" in the grid Square North 4 and East 4. Millon's designations are used in reference to individual urban sites discussed in this monograph. From 1963 - 1964, "Proyecto Teotihuacan," an excavation and restoration project of Mexico's Instituto Nacional de Anthropologia e Historia (Acosta 1964), worked on the structures along both sides of the northern section of the Miccaotli.

Sanders' "Teotihuacan Valley Project" (TVP) concentrated on the archaeological sites in the rural Valley, ultimately defining over three thousand archaeological components at nearly one thousand sites (Kolb 1979a, 1986: 157-158). Four major archaeological Periods and twenty component phases were defined, in the main, on the basis of artifactural remains. Personnel of the three projects ("Teotihuacan Mapping Project," "Proyecto Teotihuacan," and "Teotihuacan Valley Project") met to resolve problems of chronology and developed a basic ceramic typology (Kolb 1965a). The Periods and components designated by the "Teotihuacan Valley Project" were:

TF: Teotihuacan (Valley) Formative (or Preclassic) components (n = 6)
Chiconautla Phase
Cuanalan Phase
Tezoyuca-Patlachique Phase
Oxtotla Phase
Teopan Phase
Apetlac Phase

TC: Teotihuacan (Valley) Classic components (n = 6)
Miccaotli Phase
Early Tlamimilolpa Phase
Late Tlamimilolpa Phase
Early Xolalpan Phase
Late Xolalpan Phase
Metepec Phase or Terminal Classic

TT: Teotihuacan (Valley) Toltec components (n = 4);
initial Post Classic
Xometla-Oxtotipac Phase
Xometla Phase
Mazapan Phase
Atlatongo Phase

TA: Teotihuacan (Valley) Aztec components (n = 3);
final Post Classic
Aztec II or Zocango Phase
Aztec III or Chimalpa Phase
Aztec IV or Teacalco Phase

(Aztec I, Tenayuca/Culhuacan Phase is <u>not</u> found in the Teotihuacan Valley)

Modern: ca 1520 A.D. to the present (n = 1)

"Teotihuacan Valley Project" site designations used the TF, TC, TT, and TA abbreviations; for example, the Santa Maria Maquixco el Bajo Classic Period site is TC-8, <u>Teotihuacan Classic</u> (Period) site number "eight." A total of 134 Teotihuacan Classic (TC) Period sites were recognized, and these had 585 of 804 (72.76%) possible Classic component phases represented; hence, nearly all Classic sites were multicomponent. The vast majority of the 134 Classic sites also had Formative, Toltec, and/or Aztec components represented as well (2,278 potential, but 1,213 actual component phase manifestations, or 53.25%). One significant problem was to define where the urban center (TC-1) boundary "ended" and where the suburban and rural sites "began."

My particular research since 1962 sought to elaborate the Classic Teotihuacan Period artifact types and assemblages (Kolb 1965a, 1965b, 1970, 1973a, 1973b), paleopathology (Kolb 1972a, Kolb and Bilharz 1972), settlement patterns (1979a), absolute and relative chronologies (1979b), demography (1985), and commercial routes (1986). I especially have been concerned with the Teotihuacan ceramic assemblages (Kolb 1965a, 1965b), notably "Thin Orange" Ware (1973c, 1977, 1982, 1984a, 1986), "Granular Wares" (1984b), and "Copoid Wares" (1987a, 1987b).

A preliminary version of portions of this monograph, especially parts of Chapters Six and Ten, was presented as a paper, "The Old Shell Game: A Mesoamerican Trade Network: (Kolb 1973a) at the Society for American Archaeology Annual Meeting held in San Francisco, California in 1973. The original paper has been significantly modified and elaborated since, resulting in the current monograph. Dr. Anthony R. Hands, General Editor of the <u>B</u>.<u>A</u>.<u>R</u>. (<u>British Archaeological Reports</u>), suggested to me that my original title "The Old Shell Game" would probably not be understood by some readers, since the term "shell game" is a play on words. In this monograph, I detail the role of Gulf Coast (Caribbean Marine Faunal Province) and Pacific Coast (Panamanian Marine Faunal Province) marine shell importation into the Teotihuacan polity which, in turn, controlled the sources of and exportation of obsidian resources, and also exported finished products and "religious ideology." Shell importation from the coasts into the Meseta Central and Valley of Oaxaca has occurred since the Early Preclassic and continued into the Postclassic Periods as well.

The "shell game," per se, was a gambling game derived from the earlier "thimblerig" in which a person by "sleight-of-hand" manipulates a pellet (pea or bean) and three cup-like objects (walnut shells) so that the spectators can seldom know surely under which "cup" the pellet rests. The manipulator invites bets on the location of the pellet, but the "shell game" is so often played dishonestly that its name has become synonymous with chicanery or negative reciprocity, whereby the manipulator nearly always "wins." Hence, the manipulator (the Teotihuacan polity) would "get the better of the deal" from the spectator (shell supplier), in this instance by obtaining ideologically and economically important marine mollusks in "exchange" for "religious ideology" and obsidian blanks and blades. As a number of investigators have noted, the control Teotihuacan was able to exert in Mesoamerica because of its monopoly on obsidian sources and production, led to the emergence of formal exchange networks, inter-regional commerce, and professional merchant groups (Charlton 1978, 1979, 1984; Hirth 1984; Hirth and Villaseñor 1981; Kolb 1979a, 1986; Millon 1981; Santley 1977, 1984; Santley et al 1984, 1985).

Chapter One of this monograph has background material on the Classic Teotihuacan city, "state," and "empire," and briefly relates the roles of merchants and the control of obsidian sources and production for export. In Chapter Two, I delineate the Marine Faunal Provinces and three biotic families of marine mollusks -- Spondilidae (marine bivalves or clams), and Strombidae and Fasciolariidae (both gastropods or marine snails). Genera and species are also detailed. Chapter Three relates marine mollusk representations at Teotihuacan urban and rural sites, especially in mural art and lithic sculpture, but also in ceramic artifacts. In Chapter Four, mollusk specimens are reported from the burials and caches at urban center sites (Tlamimilolpa, Tlajinga 33, Xolalpan, Zacuala Palace, Tetitla, La Ventilla B, etc.). The Santa Maria Maquixco el Bajo (TC-8) site excavations, prehistory, Colonial and recent history are detailed in Chapter Five, while Chapter Six elaborates the genera and species of marine shell and cultural associations at that site. In Chapter Seven, I record the distributions of major mollusks -- Spondilidae, Strombidae, and Fasciolariidae -- as reported in site excavations and surveys throughout Mesoamerica, while Chapter Eight amplifies the importance of mollusks in nutrition and in construction, and as a source of dyes, artifact raw material (such as jewelry), and the significance of gastropod shell "trumpets." Chapter Nine details the shell procurement networks and Teotihuacan contacts with the Gulf and Pacific coasts. Seven potential commercial routes for the procurement of marine shell are delineated for each coast. The final chapter presents an evaluation of the shell procurement system and suggests a commercial model. Five appendices provide details and quantification of the marine shell specimens and associated TC sites.

The text is accompanied by a "Glossary of Architectural Terms" and a "Glossary of Spanish and Nahuatl Terms." The former relates specifically to Figures 8-15 and 19-23, which are architectural plans of various sites and locations of marine shells, art, etc., while the latter glossary is especially useful in defining terms used in Chapters One and Five, and in Appendix V. The maps. plans, artifact illustrations, and photographs are all by the author. "North" arrows on the maps and plans consistently indicate Magnetic North.

I would like to acknowledge the assistance of William T. Sanders and Rene F. Millon, who cordially permitted me access to artifact collections and field notes, and I thank them for providing the following materials: My Figure 3, "Chronological Concordances for the Basin of Mexico," is derived from Sanders et al (1979) and Kolb (1979a), while Figure 7, "Urban Teotihuacan During the Xolalpan Phase," is reproduced from a 1970 uncopyrighted version of Millon's (1973:Map 1) base map. I am indebted to Harold S. Feinberg, Department of Living Invertebrates, American Museum of Natural History for his assistance in classifying the Teotihuacan Valley specimens and for stimulating discussion about marine mollusks, taxonomic problems, and clarifications of the Spondilidae. Portions of the manuscript were read by William T. Sanders, Robert S. Santley, Jeffrey R. Parsons, and George Cowgill. I am indebted to them for their comments, corrections, and advice. Drafts and the final version of this monograph were word procedded by Wendy Eidenmuller, Joy Kolb, and Mary Jeanne Weiser, and I am thankful for their skills and careful attention to detail. Dr. Anthony R. Hands was especially helpful in matters of format and illustration reproduction. I, of course, bear final responsibility for any errors in judgment and interpretation.

> Charles C. Kolb
>
> Erie, Pennsylvania, U.S.A.
>
> August 1, 1987

CHAPTER ONE:

CLASSIC PERIOD TEOTIHUACAN

Introduction to the Research

For over twenty thousand years the Basin of Mexico, a hydrographic province of nearly eight thousand square kilometers, was one of the major archaeological regions of the Mesoamerican Culture Area (Sanders 1981). (See Figures 1, 2 and 3.) The Basin was the location of the waxing and waning of two great polities, Classic Period Teotihuacan (ca 50-750 A.D.) and, subsequently, the late Postclassic Aztec/Culhua Mexica centered at Tenochtitlan (ca 1250-1520 A.D.). Both developed city-states which bore the basic characteristics of "civilization," with associated elaborations in sociopolitical, religious, and economic realms.

In this monograph I shall consider an economic aspect of Classic Teotihuacan society which affected all realms, namely, exchange networks in "luxury" or "exotic" raw materials and finished products. Among a vast array of perishable and non-perishable material culture were obsidian, ceramics, and marine shell, as well as textiles, feathers, and foodstuffs. While emphasis will be placed on the importation of marine shell into Teotihuacan society, the roles of other goods and services cannot be ignored. I shall initially present a general overview of the Teotihuacan Supra-Regional Center, and its associated "state" and incipient "empire" (Kolb 1986:171-172, 185-189). The procurement of raw materials, the production of finished products, and general mechanics of distribution will be detailed. The major marine mollusks of importance to Teotihuacan will be considered in terms of sources, species, and uses. Representation of mollusks in the mural art, lithic sculpture, and minor arts will be reviewed, and the occurrences of shells in <u>ofrendas</u> (offerings) and as mortuary goods in the urban center will be elaborated.

In a more detailed consideration of the economics of marine shell procurement, I shall suggest the role of the Santa Maria Maquixco el Bajo Classic site in the network of importation of <u>Spondylus calcifer</u> Carpenter, 1857 into the Teotihuacan polity. These mollusks were found in extraordinary quantity at this rural Teotihuacan Valley site, one of the first excavated as a part of William Sanders' "Teotihuacan Valley Project," the initial segment of the long-range "Basin of Mexico Survey

FIGURE 1: THE MESOAMERICAN CULTURE AREA

(after J. R. Parsons 1971:3, Map 1; Sanders and Price 1968:8, Fig. 1; Kolb 1979a:3, Map 1)

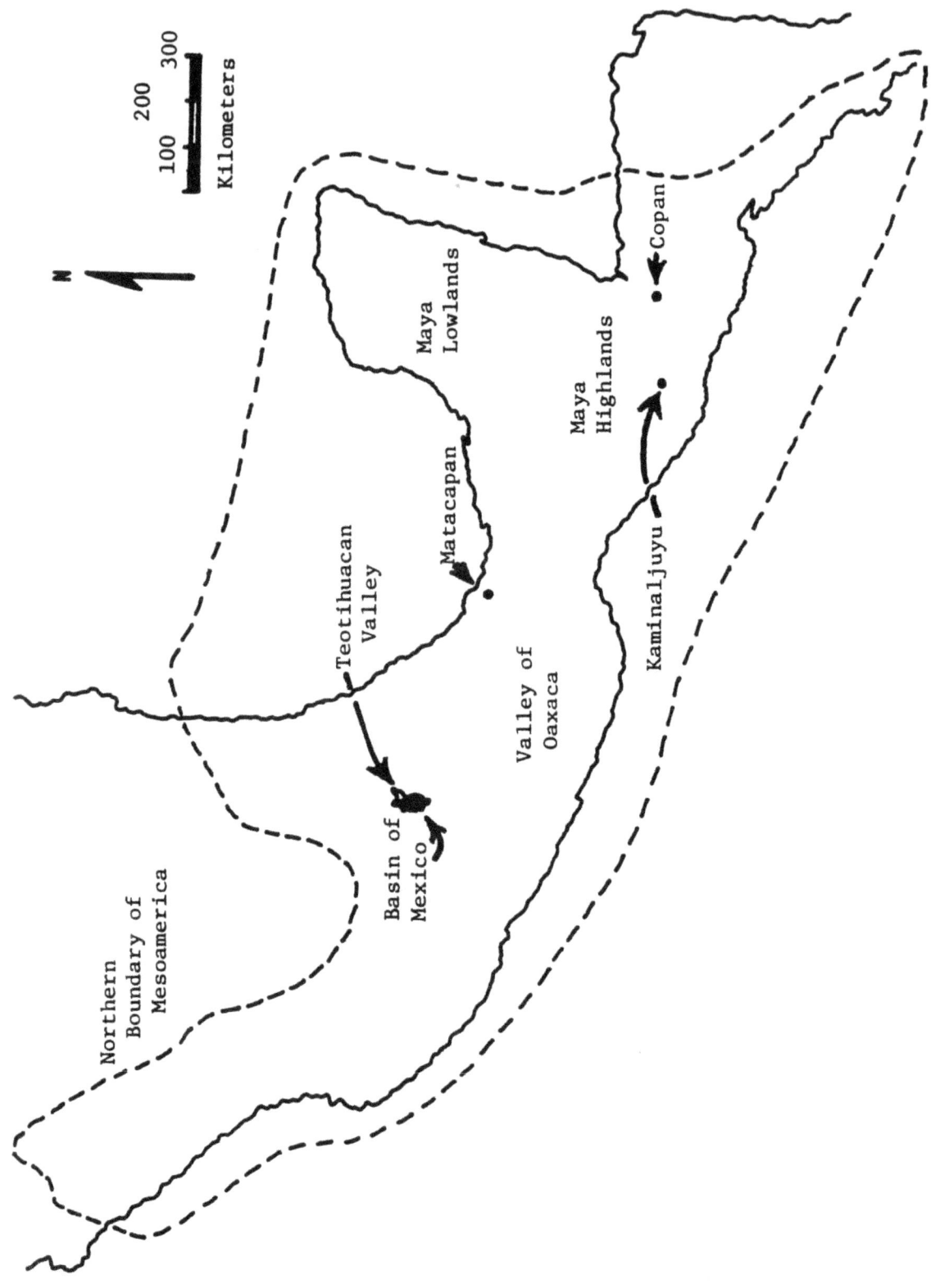

FIGURE 2: THE BASIN OF MEXICO

(after J. R. Parsons 1971:4, Map 2; Kolb 1979a:9, Map 2, 1986:161, Fig. 1; Sanders et al 1979:Separate Map 2)

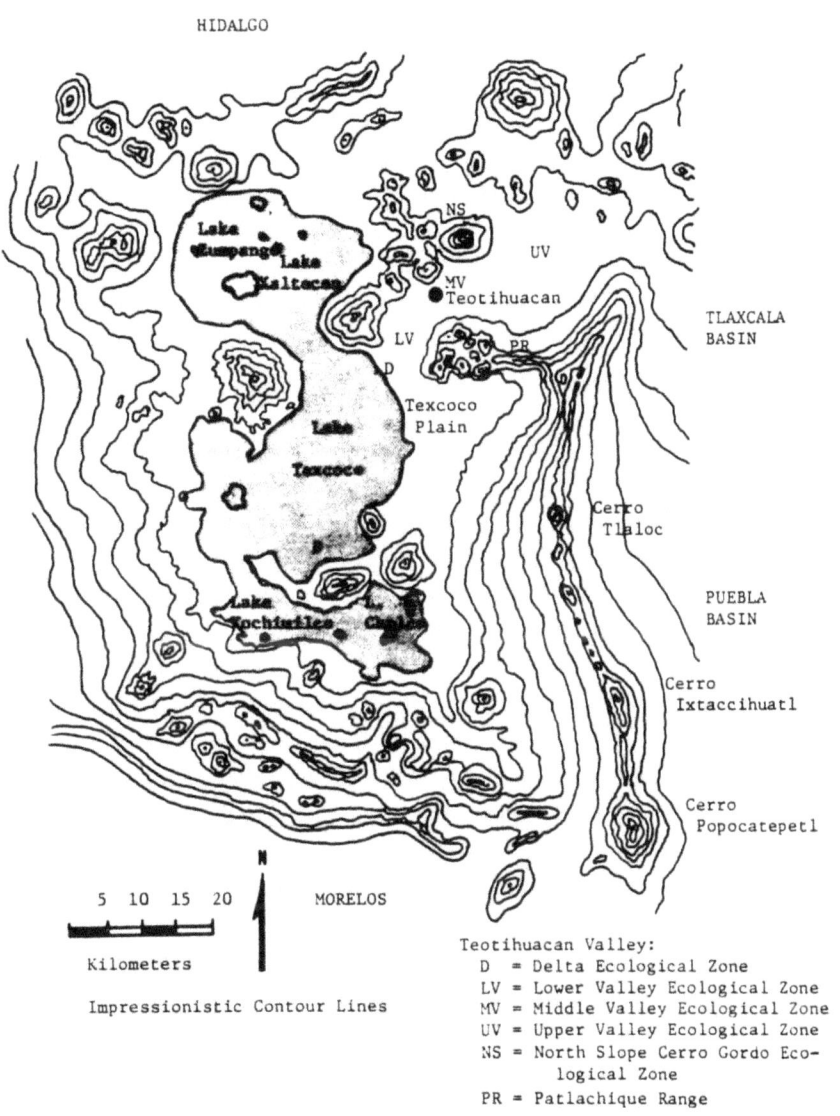

Teotihuacan Valley:
D = Delta Ecological Zone
LV = Lower Valley Ecological Zone
MV = Middle Valley Ecological Zone
UV = Upper Valley Ecological Zone
NS = North Slope Cerro Gordo Ecological Zone
PR = Patlachique Range

FIGURE 3: CHRONOLOGICAL CONCORDANCES FOR THE BASIN OF MEXICO

(after Kolb 1979a:65-66, Table 5, 1979b; Sanders et al 1979:93, Table 5.1)

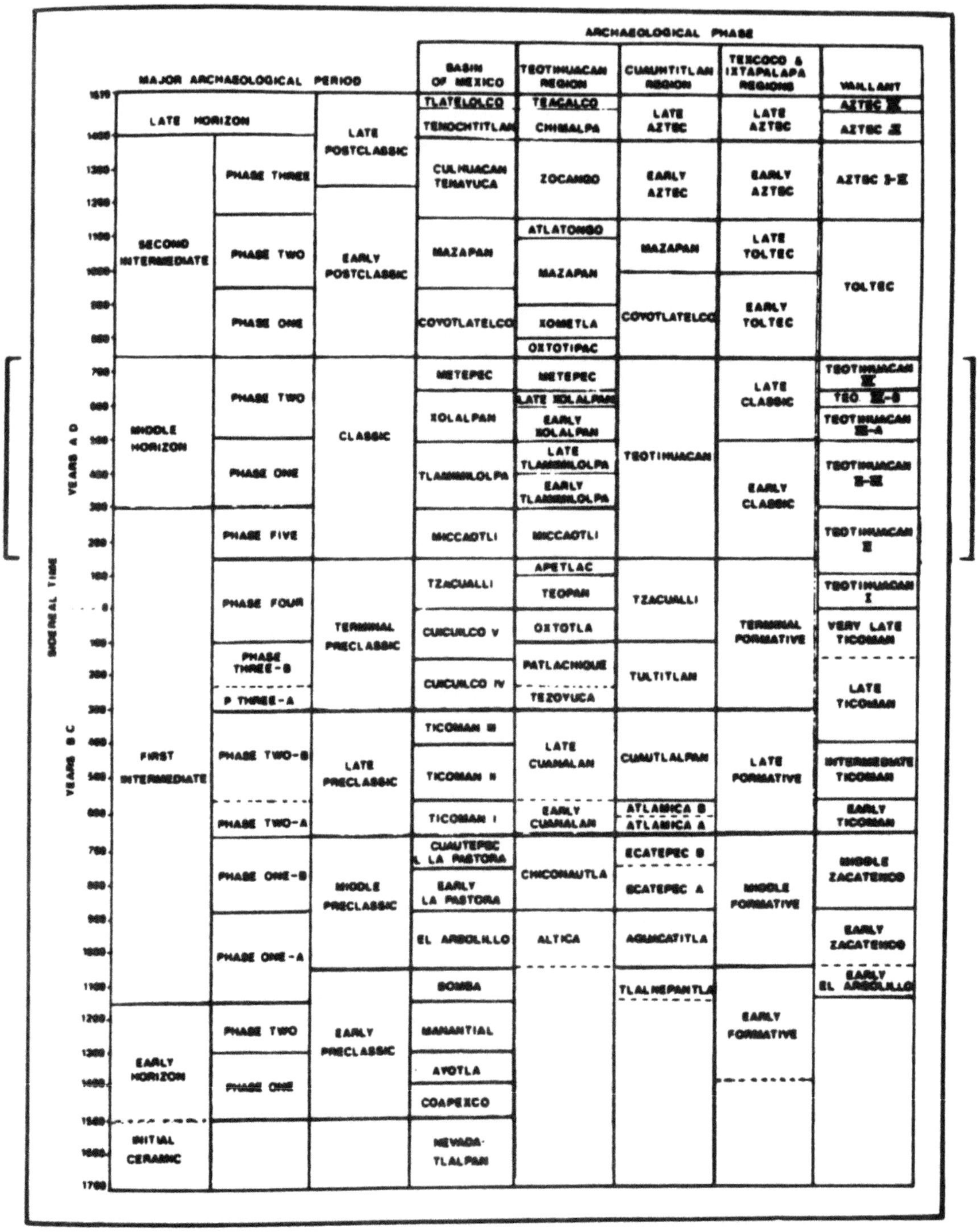

	MAJOR ARCHAEOLOGICAL PERIOD		ARCHAEOLOGICAL PHASE				
			BASIN OF MEXICO	TEOTIHUACAN REGION	CUAUHTITLAN REGION	TEXCOCO & IXTAPALAPA REGIONS	VAILLANT
1519	LATE HORIZON	LATE POSTCLASSIC	TLATELOLCO	TEACALCO	LATE AZTEC	LATE AZTEC	AZTEC IV
1400			TENOCHTITLAN	CHIMALPA			AZTEC III
1300	SECOND INTERMEDIATE	PHASE THREE		CULHUACAN TENAYUCA	ZOCANGO	EARLY AZTEC	EARLY AZTEC	AZTEC I-II
1200							
1100		PHASE TWO	EARLY POSTCLASSIC	MAZAPAN	ATLATONGO	MAZAPAN	LATE TOLTEC	TOLTEC
1000					MAZAPAN			
900		PHASE ONE		COYOTLATELCO	XOMETLA	COYOTLATELCO	EARLY TOLTEC	
800					OXTOTIPAC			
700	MIDDLE HORIZON	PHASE TWO	CLASSIC	METEPEC	METEPEC	TEOTIHUACAN	LATE CLASSIC	TEOTIHUACAN IV
600				XOLALPAN	LATE XOLALPAN			TEO IV-A
500					EARLY XOLALPAN			TEOTIHUACAN III-A
400		PHASE ONE		TLAMIMILOLPA	LATE TLAMIMILOLPA		EARLY CLASSIC	TEOTIHUACAN II-III
300					EARLY TLAMIMILOLPA			
200		PHASE FIVE		MICCAOTLI	MICCAOTLI			TEOTIHUACAN II
100					APETLAC			TEOTIHUACAN I
0		PHASE FOUR	TERMINAL PRECLASSIC	TZACUALLI	TEOPAN	TZACUALLI	TERMINAL FORMATIVE	VERY LATE TICOMAN
100				CUICUILCO V	OXTOTLA			
200	FIRST INTERMEDIATE	PHASE THREE-B			PATLACHIQUE	TULTITLAN		LATE TICOMAN
300		P THREE-A		CUICUILCO IV	TEZOYUCA			
400		PHASE TWO-B	LATE PRECLASSIC	TICOMAN III	LATE CUANALAN	CUAUHTLALPAN	LATE FORMATIVE	INTERMEDIATE TICOMAN
500				TICOMAN II				
600		PHASE TWO-A		TICOMAN I	EARLY CUANALAN	ATLAMICA B / ATLAMICA A		EARLY TICOMAN
700		PHASE ONE-B	MIDDLE PRECLASSIC	CUAUTEPEC / LA PASTORA	CHICONAUTLA	ECATEPEC B	MIDDLE FORMATIVE	MIDDLE ZACATENCO
800				EARLY LA PASTORA		ECATEPEC A		
900		PHASE ONE-A		EL ARBOLILLO	ALTICA	AGUACATITLA		EARLY ZACATENCO
1100				BOMBA		TLALNEPANTLA		EARLY EL ARBOLILLO
1200	EARLY HORIZON	PHASE TWO	EARLY PRECLASSIC	MANANTIAL			EARLY FORMATIVE	
1300								
1400		PHASE ONE		AYOTLA				
				COAPEXCO				
1500		INITIAL CERAMIC		NEVADA TLALPAN				
1700								

Project" (Sanders, Parsons, and Santley 1979). Lastly, with emphasis on the Meseta Central, I shall present a model for shell exchange. The Mesoamerican shell procurement network, which developed in Early Preclassic times (Pires-Ferreira 1978:83-84), was enhanced during the Classic Period in the Teotihuacan-influenced Meseta Central and in the Classic Maya regions of the Yucatan Lowlands and Guatemalan Highlands. Peoples far removed from the Pacific and Gulf of Mexico-Caribbean coasts sought marine conch and bivalve shells for ritual, ceremonial, and ornamental purposes.

The Teotihuacan polity intensified and elaborated the shell procurement network into a significant aspect of their commercial network and "empire" expansion which touched both coasts in order to secure sufficient quantities of mollusk to enhance the "elite" (high status) members of Teotihuacan society and, in turn, the probable theocratic polity itself. Teotihuacan's ability to control obsidian sources, production, and subsequent distribution (Charlton 1978, 1979, 1984; Hirth 1984a; Santley 1977, 1984; Spence 1984) were a part of the shell procurement network. Obsidian artifacts were a major Teotihuacan export, while the importation of <u>Alpha</u> "Thin Orange" Ware (Kolb 1973c, 1982, 1984a, 1986) and "Granular Ware(s)" (Kolb 1984b), along with marine shells were possibly under "state" control (Kolb 1973a).

Teotihuacan: City, "State," and "Empire"

The City and Its Craft Specializations

The Classic Period metropolis of Teotihuacan was located in the middle segment of the valley of the same name in the northeastern quadrant of the Basin of Mexico (Figures 2, 3, 4 and 5). The Teotihuacan Valley, a physiographic zone of approximately six hundred square kilometers (including adjunct zones such as the North Tributary Valley and Cerro Gordo North Slope areas), opened into the Basin proper to the southwest, while, to the east, it was separated by a low range of foothills from the Basin of Puebla-Tlaxcala. The Sierra de Patlachique, Sierra Nevada with Cerro Tlaloc, and Cerros Ixtaccihuatl and Popocatepetl were a north to south barrier separating the Basins of Mexico and Puebla-Tlaxcala. (See Figure 6.)

Urban Teotihuacan was a spatially large and densely populated Supra-Regional Center during the Classic and encompassed a maximal area of 122.5 square kilometers

FIGURE 4: THE TEOTIHUACAN VALLEY DURING THE CLASSIC PERIOD

(after Sanders 1965:Fig. 6.; Kolb 1979a:211, Map 4)

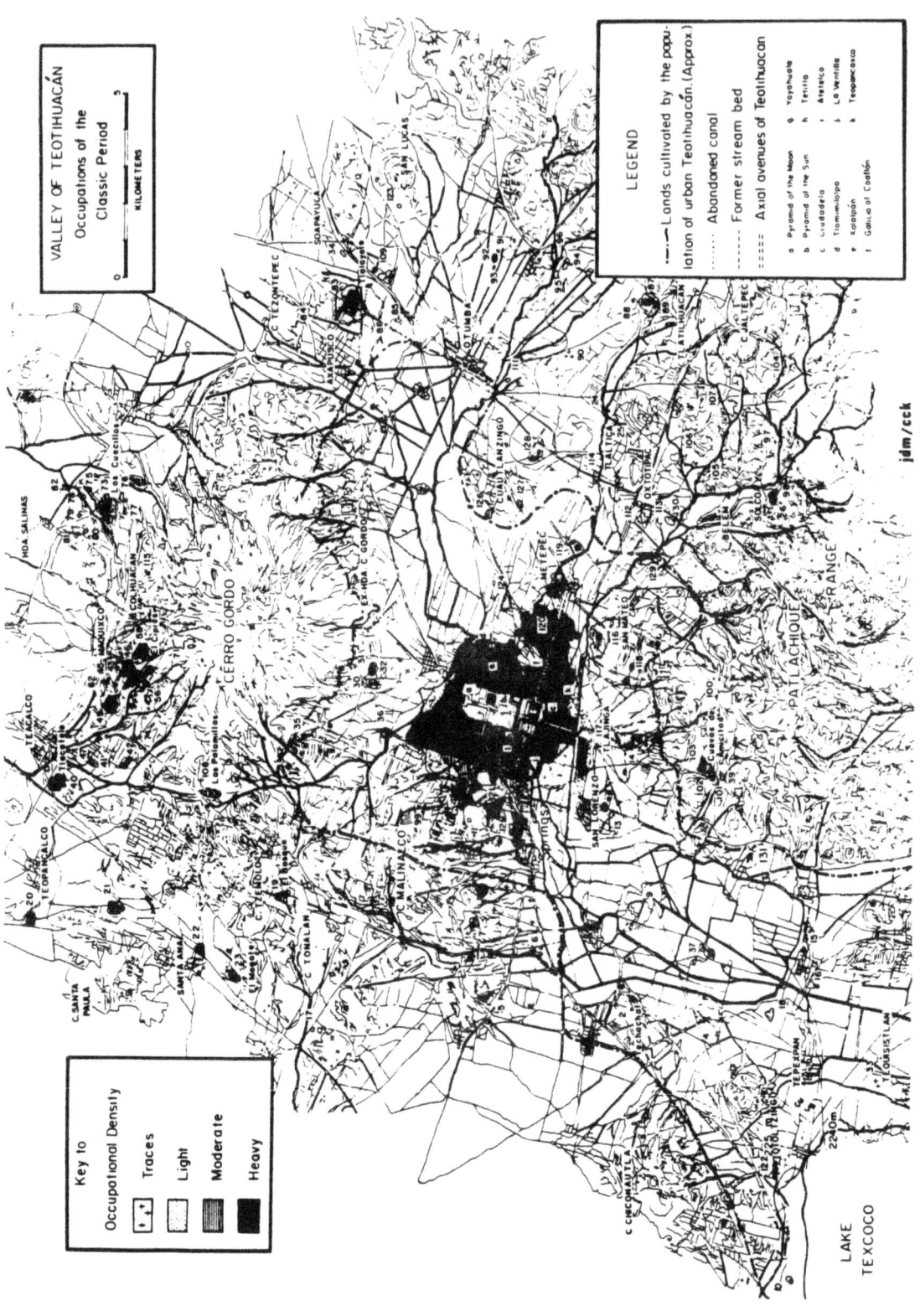

FIGURE 5: URBAN TEOTIHUACAN DURING THE XOLALPAN PHASE
(ca 600 A.D.)

(after Millon 1973:Map 1; Kolb 1979a:213, Map 5)

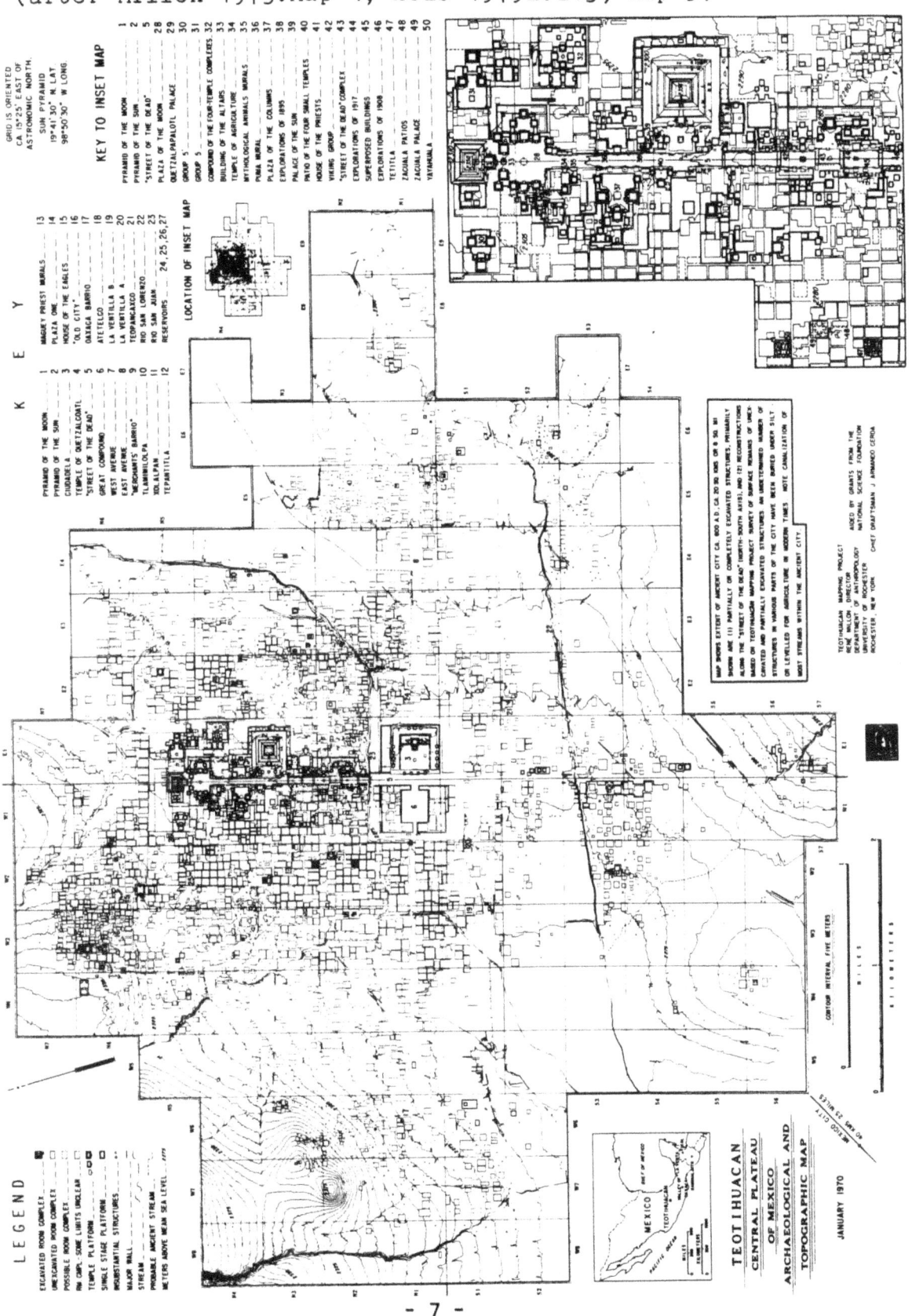

FIGURE 6: THE BASINS OF MEXICO, TLAXCALA AND PUEBLA:
           MAJOR COMMERCIAL ROUTES

(after Kolb 1986:178, Fig. 2)

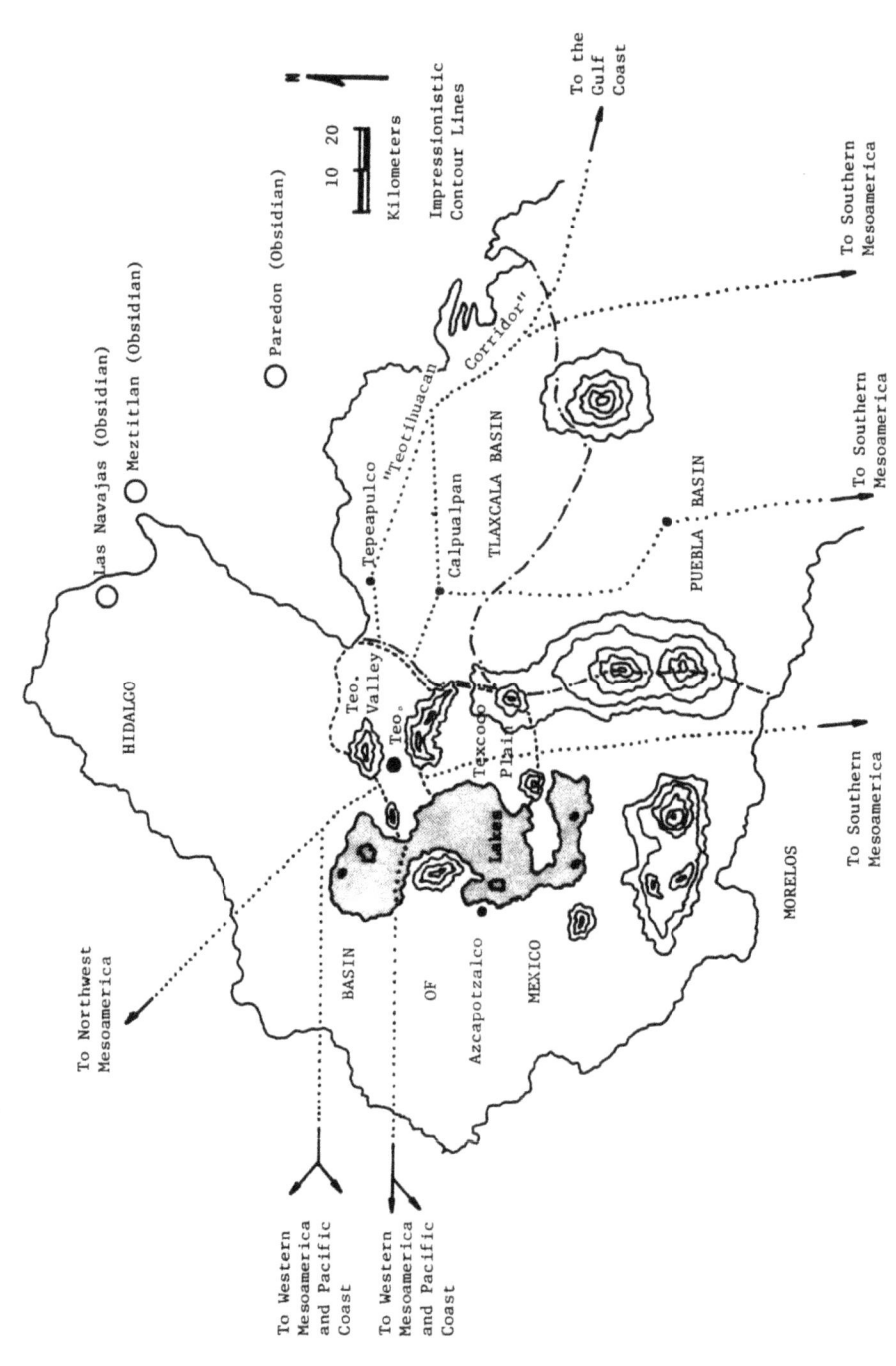

with a population estimated at a minimum of 85,000 but probably ranging from 125-150,000, or even up to 200,000 as a possible maximum (Millon 1970:1077, 1976:107, 1981:208; Sanders et al 1979:109; Sanders 1981:176). At its spatial and demographic peaks during the Xolalpan phase (ca 450-650 A.D.) the urban center had densities as high as seven thousand inhabitants per square kilometer. The Supra-Regional Center contained monumental religious, administrative, and commercial architecture (the Pyramids of the Sun and Moon, Ciudadela, and Great Compound), with residential areas reflecting different social status and areas of craft specializations.

The urban center was bisected from north to south by a wide avenue called the Miccaotli ("Street of the Dead," its Aztec name), which provided a city planning orientation of 15° 25' east of magnetic north. The Pyramid of the Moon (derived from the Spanish "Piramide de la Luna") provided a northern terminus for this north-south avenue. To the south, on the east side of the Miccaotli, was the Pyramid of the Sun ("Piramide del Sol"), the largest architectonic feature of the city. Further to the south, also on the east side of the Miccaotli, was the Citadel ("Ciudadela") with an older Temple of Quetzalcoatl and less ornate "replacement" temple facade (Plataforma Adosada) situated within the rectilinear precinct. At the rear (east) of the Ciudadela another major avenue, 3.5+ kilometers in length, the East or Eastern Avenue, was constructed at a "right angle" (91°) to the Miccaotli.

Facing the Ciudadela immediately to the west across the Miccaotli was the Great Compound (Gran Conjunto Recinto Mercado), an architectural counterpart to the Citadel. At the rear (west) of the Great Compound another major avenue, the West or Western Avenue, continued west at a right angle to the compound for approximately 3.0 kilometers. In effect these three avenues formed a cruciform grid, dividing the city into distinct quadrants (Kolb 1979a:223). While the Pyramids of the Moon and Sun and nearly one hundred small temple platforms flanking the two kilometer long Miccaotli were interpreted as religious structures, the Ciudadela was more likely an administrative center of the probable Teotihuacan "theocratic" polity. Its counterpart, the Great Compound, Site 1:N1W1 (a site abbreviation designated by the Teotihuacan Mapping Project), was the economic precinct of urban Teotihuacan and included the largest marketplace, and bureaucratic and storage (bodega) facilities (Millon 1973).

A total of 5,047 "sites" were identified within the metropolis, including over two hundred "domestic" ceramic workshops, nearly two hundred ceramic figurine workshops, and approximately four hundred obsidian-working sites (Millon 1973:45-46, 1976:232-233, 1981:223-225; Barbour 1976:173-227; Ester 1976:40-42; Krotser and Rattray 1980:92-99; Spence 1984:100-108; Kolb 1987a, 1987b). Lesser numbers of ground stone (basalt, slate, and volcanic scoria (tezontli) workshops, other lapidary work areas, and shell artifact workshops were delineated primarily on the basis of manufacturing debris. Approximately 2,600 one-story residential compounds, interpreted as "apartments" inhabited by a number of related households, probably by lineages and/or clans (calpulli) which housed from sixty to one hundred people each (Millon 1976:215-216, 233; 1981:206). Ethnic barrios or "neighborhoods," often associated with craft specialization, were discerned. Oaxacan, Huastec, and apparently Maya enclaves were also identified. The craftspersons and their immediate families exceeded 30,000 individuals (Millon 1976:233, 1981:210).

In addition to the more obvious ceramic and lithic workshops, the urban center also undoubtedly had craftspersons working in various organic raw materials. Less easily identified were the workshops of artisans who manufactured cotton and maguey (Agave americana) textiles, basketry, mats (petates), wood, leather, and featherwork. Tropical bird feathers, especially those of the quetzal, were woven into capes and garments, and made into insignia and badges of rank, to judge from depictions in Teotihuacan mural art (Villagra Caleti 1971; Miller 1973, 1978). Amantecatl, feather workers documented during the time of the Postclassic Aztec state, were extremely skilled artisans who occupied a status position just below that of the elite-nobles, bureaucrats, and military officials (Sahagun 1959:83-85, 1961:25). Seler (1915:453-454) believed that the amantecatl owed their cultural origins to Classic Teotihuacan.

Merchants: "Foreign" and Domestic

The roles and status(es) of mechants are inferred from artifact assemblages and mural art. Tetitla (Site 1:N2W2) may have been the residence and possibly the workshops of feather merchants, while traders in maize might have occupied the Zacuala Palace (Site 3:N2W2), and ceramic artisans lived and worked at Teopancaxco (Site 1:S2E2) and in the Tlajinga barrio (especially Site 33:S3W1; but also other loci in S3W1, S3W2, S4W1, and S4W2). Millon (1967b:42-44, 1973:40-42; 1976:233-234; 1981:222-223, 226) identified a "merchants' barrio," with concentrations of Veracruz Gulf Coast and Lowland Maya ceramics (sites named Xocotitla, 3:N3E4, and Mezquititla 8:N4E4). The "merchants' barrio" did not appear to have been occupied by foreigners, but probably served as a combination bodega (storehouse) and possible "customs house" (Kolb 1986:172). Millon (1981:226) reported quantities of a white slipped "Fine Orange" ware which came from Matacapan and the Tuxtlas region of the southern Veracruz coast. A multi-room adobe structure in 8:N4E4 was excavated (TE-4) and produced concentrations of Gulf Coast and Lowland Maya ceramics. The "Maya" sherds were not further defined.

Millon (1967:42-43, 1973:41-42, 1976:233-234, 1981:222-223, 241; Spence 1976:130) defined the "Oaxaca barrio, Tlailotlacan, (Site 7:N1W6), identified on the basis of fine and utilitarian Oaxaca wares and "Oaxaca-like" pottery made from local Teotihuacan clays. Excavation (TE-3) in this barrio yielded two urns, one dated to the Monte Alban Epoch of Transition and the second to Monte Alban IIIA. The former "heirloom" urn was associated with an extended position human burial assumed to be Oaxacan, since Teotihuacan burials were nearly always in flexed positions, but extended burials were common in Oaxaca (Spence 1971:114, 1976:145). A seventh century A.D. stone-lined Oaxaca style tomb had an associated stela carved in Monte Alban style which had a Oaxacan glyph depicting "movement" and below it the number "nine" represented by a bar designating five and four dots beneath designating the invididual numbers "one." The Monte Alban IIIA urn (dated ca 250-450 A.D.), the tomb and the glyph styles, and ceramic evidence suggested a probable three hundred year occupation "by Oaxaqueños" from ca 400-750 A.D. (Millon 1973:42).

It was inferred that the barrio inhabitants were traders from Oaxaca who retained their own social and cult traditions, or were Teotihuacanos who adopted some Oaxacan cultural traits. I have postulated that the barrio might not have been the residence of Oaxaqueños but simply the bodega (warehouse) where imports were kept prior to transshipment to Teotihuacan's marketplaces, including the Great Compound, but the Oaxaca barrio may also have functioned as a "customs house" or served as the residence of Teotihuacanos engaged in "long-distance" trade (Kolb 1986:181-195).

Spence (1971:131-132) reported the possible use of Gulf Coast and Oaxacan ceramics as "exotic goods used in mortuary activities" associated with the N4W3 Crematory Area (Sites 24, 29, 30, 31, and 32:N4W3). However only five Gulf Coast and six Oaxacan sherds were among the over seven hundred pottery fragments in the contexts of twenty-eight identifiable human cremations (1971:117, 120, 128).

The "merchants'" and "Oaxaca" barrios were probably not the only ethnic enclaves in the urban center. A Huastec enclave (Square N2W6), possibly associated with a two-tiered round temple (69:N2W6), was located north of the "Oaxaca barrio" (Millon 1973:35). The La Ventilla A compound (Site 1:S1W3) may have been the residence of craft specialists in stone and shell who had emigrated from the Veracruz coast (Piña Chan 1963:52). A human burial with abundant shell and some central Veracruz ceramics suggested to Piña Chan that commercial contacts with the Gulf Coast continued during Teotihuacan's apogee, but Millon (1976:234) discounted this proposition.

No enclaves of West Mexican peoples have been identified in the urban center or the Teotihuacan Valley, but these probably existed. Specific Maya residences likewise have not been differentiated although the Teotihuacan-Kaminaljuyu connection would suggest the possibility of a separate Maya enclave. However, in the commercial route to the Gulf Coast known as the "Teotihuacan Corridor," one community was settled by peoples from El Tajin ca 600 A.D. (Garcia Cook and Trejo 1977, Garcia Cook 1981:266). This route would also lead to the central Gulf Coast and the site of Matacapan located on the coastal plain between Lake Catemaco and the slopes of San Martin Tuxtla volcano (Coe 1965:683; Millon 1981:226; Kolb 1986:195; Santley et al 1984, 1985). This site of over seventy mounds had Teotihuacan style talud and talero architecture, an axial orientation slightly east of north, and "Thin Orange" ceramics. It certainly had Teotihuacan influence and may have been a provicial Center of the Teotihuacan "empire" or at least was similar to Kaminaljuyu, a Teotihuacan "outpost."

Teotihuacan Obsidian and Other Resources

Central to the political and commercial sucess of the Teotihuacan polity was its ability to control the obsidian sources, production, and distribution within the "state" and "empire," and beyond. Grey obsidian sources at Otumba, immediately east of the urban center in the Teotihuacan Valley, fine green obsidian from Cerro de las Navajas, Paredon, and Meztitlan Valley sources, all less than sixty kilometers to the northeast of Teotihuacan, were within Teotihuacan's jurisdiction (Charlton 1978:1229-1230, 1984:19-20; Millon 1976:231, 1981:217, 223-225; Spence 1984:91, 94; Kolb 1986:173-174). A coarse green obsidian from the Pizarin source (Charlton 1978:1229-1230; Spence 1984:94) was exploited to a lesser extent. Obsidian was processed in specialized workshops at Teotihuacan especially clustered in areas adjoining the Pyramid of the Moon, the Ciudadela, and Great Compound. The nearly four hundred workshops produced implements for the urban center and rural sites, and made pointed base "blanks" and prismatic blades for export throughout the "empire" (Ester 1976:101-109).

The site of Tepeapulco, Hidalgo, which was under Teotihuacan control, imported raw obsidian from the Otumba, Navajas, and Paredon sources (Charlton 1978:1128, 1984:20-22). Large workshops at this site produced implements for export to the south and east rather than for Teotihuacan or other sites in the Basin of Mexico. By Early Xolalpan times, the Teotihuacan polity held a monopoly on the production and distribution of obsidian. An obvious limited analogy may be drawn to OPEC's (Organization of Petroleum Exporting Countries) control of major sources of crude petroleum destined for Japanese and Western European consumers 1973-1984. Teotihuacan's obsidian monopoly was more pervasive since Mesoamerican consumers did not have access to alternative sources of obsidian cutting tools as modern petroleum consumers have to crude oil. In addition, Teotihuacan's commercial activities were linked to religious ideology and, to some extent, its military power.

Late in its history, ca 600 A.D. the Teotihuacan "state" was apparently involved in the exploitation of cinnabar from mines in the Sierra de Queretaro some two hundred kilometers north of the Basin of Mexico (Millon 1976:232, 1981:227). As a ground mineral this orange-red pigment was used in mortuary rituals (Spence 1971:90-97, 1976:133-134, 137-138) and probably other secular and religious activities, but was also employed in decorating champleve ceramics and ceramic figurines (Kolb 1965a, 1973b). Approximately three hundred kilometers to the north in the Chalchihuites region of Zacatecas, Teotihuacan may have been involved in the mining of several types of raw materials from ca 250-500 A.D. (Millon 1981:227). Chief among these was hematite, a red pigment used at Teotihuacan in architectural decoration (primarily wall painting), mural painting, and ceramic decoration. Cinnabar, flint, "white jade," and "blue-green mineral stones" were also mined. Turquoise from the Cerrillos region of New Mexico may have been channeled into the Teotihuacan polity through Chalchihuites (Millon 1981:227-228).

The Urban Social System

The Supra-Regional center was a religious complex and the focus of the "state" religion, but was also a secular city with political-theocratic, social, and economic functions. Based on architectural styles seen in residences, mural art, and domestic artifact assemblages, levels of socioeconomic subordination existed. Millon (1976:227-228, 1981:212-214) has argued that there was evidence for six or more social levels, with a priestly and administrative hierarchy at the top, living in apartment compounds in the Ciudadela. A second stratum inhabited apartment compounds of nearby temple complexes and residences in the Great Compound. Three intermediate status levels lived in various residential compounds (Atetelco, La Ventilla A and B, Teopancaxco, Tepantitla, Tetitla, Tlamimilolpa, Xolalpan, Yayahuala, and Zacuala), while the few thousand people of the lowest status lived in insubstantial one- and two-room adobe structures (chozas) throughout the city. The intermediate status persons comprised the demographically largest social units of the urban center. Examples of residences of these three statuses from highest to lowest were Zacuala Palace, Teopancaxco, and Xolalpan (Millon 1976:227). La Ventilla B, Tlamimilolpa and Tlajinga 33 (Storey and Widmer 1982) were lower status habitations. The several thousand members of the first and second social strata undoubtedly controlled Teotihuacan's economic as well as sociopolitical and religious life.

In addition to the craft specialists, inhabitants of the ethnic conclaves, and merchants who occupied the intermediate (middle three) social strata in the urban center, a variety of other specialists would also have been included. Except for inferences from mural art and the artifact assemblages in the apartment compounds, we know little of the other artisans who undoubtedly were important to the maintenance of the probable theocratic polity. Ceramic figurines suggest ritual specialists such as healers, curers, or "medical" practitioners (Sahagun 1961:139-164; Barbour 1976:172-173, 179, 182, 184, 196, 196, 197, 198, 216, 222). Postclassic Aztec analogy, ca 1519, suggested that musicians, singers, dancers, poets, actors, magicians, acrobats, athletes, ball players, teachers, vendors, and cooks or chefs would have probably been practicing their trades (Sahagun 1961:25-44, Borhegyi 1971:84, Millon 1976:235-236).

Likewise, urban planners, architects, stone masons, carpenters, sculptors, mural painters, and others contributed their talents to the construction and maintenance of the urban center. Millon (1976:236) postulated that the city also had prostitutes, thieves, con men, and other such individuals. These, in turn might require the presence of a religious or police force with coercive powers, and a judicial system as well. The diversity of occupations was probably close to that of the Postclassic Aztec city of Tenochtitlan, ca 1519 (Sahagun 1954, 1959, 1961). Merchants, both wholesalers and retailers (Sahagun 1959:1-68), vendors (1961:63-95), and artisans who worked in precious stones and feathers (1959:69-97, 1961:25-36) comprised a significant part of the economic life of Tenochtitlan, and undoubtedly Classic Teotihuacan as well.

The Rural and Provincial Regions

In the rural areas of the Teotihuacan Valley during the demographic apogee of the Xolalpan phase, 109 sites have been identified (Kolb 1979a:219-245, 287, Table 59, 560-563). There are a total of 134 Classic Teotihuacan Period sites in the Valley, hence, twenty-five were not occupied during the Xolalpan phase. These 109 sites included three Provincial Centers (sites TC-40, TC-73, and TC-83), six Large Nucleated Villages, twenty-one Small Nucleated Villages, one Small Dispersed Village, thirty-eight Hamlets, one Hamlet with Special Activity Loci, twenty-nine Isolated Residences (chozas), four Special Activity Loci, and six Ceremonial Precincts. The maximum population of these rural sites was 21,800-22,700 (Kolb 1979a:296-298; Sanders et al 1979:393-395, Sanders 1981:176-177).

The Teotihuacan "state," centered at the Supra-Regional Center (a "primate" city), encompassed the Teotihuacan Valley and most, if not all, of the Basin of Mexico and included a total of 253 Classic sites of the Xolalpan phase (Sanders et al 1979:108-129, Sanders 1981:175-180). In addition to the "primate" city, nine Provincial Centers, seventeen Large Villages (fifteen Nucleated, two Dispersed), seventy-seven Small Villages (fifty-five Nucleated, twenty-two Dispersed), and 149 Hamlets (including chozas) were discerned. Among the Provincial Centers were Azcapotzalco, located on the west side of Lake Texcoco, and Portezuelo in the southeastern Basin (Vaillant 1934; Hicks and Nicholson 1964:500; J. Parsons 1971:196-198; J. Parsons et al 1982; Kolb 1986:172, 178, 179). Eighteen non-residential sites, nine Isolated Ceremonial Precincts, two Large Ceremonial Precincts, four indeterminate sites, and three Quarries (two obsidian, one gravel) plus four Special Activity Loci (Teotihuacan Valley) were also located. No ceramic, pottery figurine, lapidary, or shell artifact workshops have been identified in any excavated or surveyed sites outside of the "primate" center.

The eight thousand square kilometer Basin of Mexico constituted the core of the Teotihuacan "state," but the process of expansion during the Early Tlamimilolpa phase (ca 300-400 A.D.) led to the formation of a probable Teotihuacan "empire." The polity expansion was predominantly for commercial reasons so that foodstuffs (both "basic" and "exotic"), raw materials, and luxury goods ("Thin Orange" ceramics, for example) and possibly services were imported from within the state and the hinterlands that formed the "empire" (Kolb 1984a:209-210). The hinterlands provided regional markets for the distribution of Teotihuacan's urban craft products, especially obsidian "blanks," cylindrical cores and prismatic blades, and ceramics.

To the south of the Basin, the Rio Amatzinac Valley of eastern Morelos had 121 Classic Teotihuacan sites dating to the Xolalpan phase, and the Coatlan region of western Morelos had thirty-nine sites (Hirth and Villaseñor 1981:140-143, Kolb 1984b:35). The former region may have been a cotton-producing area for Teotihuacan. The Tula region in Hidalgo to the north of the Basin of Mexico was apparently a source of construction materials, lime, and probably timber for urban Teotihuacan (Cobean 1978:84-85). At least fourteen Classic sites are known for the region, including a Provincial Center.

Immediately east of the Teotihuacan Valley lay the Plains of Apan in southeastern Hidalgo and northern Tlaxcala, where five Provincial Centers were defined (Garcia Cook and Trejo 1977:57-68). The Puebla-Tlaxcala region had sixty Classic sites, while the Tlaxcalan Altiplano had 445 (Kolb 1984b:7). A major commerical route, the "Teotihuacan Corridor" led from the Teotihuacan Valley east through the Tenenyecac region ultimately to the Gulf Coast and southern Puebla, bypassing Cholula territory (Garcia Cook and Trejo 1977:66-68; Charlton 1978:1235-1236, 1984:18-29; Kolb 1979a:248-249, 1986:176-181). I have defined two other commercial routes, an "Eastern Basin of Mexico Route (Route One)" linking urban Teotihuacan and Morelos, and the "Western Basin of Puebla-Tlaxcala Route (Route Two)," linking Teotihuacan, Cholula and southern Puebla (Kolb 1984a:217-219, 1986:179-181). Both routes were involved in the importation of "Thin Orange" ware and the export of Teotihuacan obsidian. Litvak King (1978:120) has postulated a "Rio Balsas route" leading from Guerrero through western Morelos into the Basin of Mexico. In the latter case, the assumed import into urban Teotihuacan was cacao, with Teotihuacan obsidian as a probable export.

The "State" and "Empire"

The "primate" center was the political and economic hub of a commercially expansive "empire." Polity, economy, and religious ideologies were inextricably linked within the "state" and to a lesser extent in the hinterlands comprising the "empire." These regions encompassed some 25,000 square kilometers, an area comparable in size to the Commonwealth of Massachusetts or the nation of Belgium (Millon 1981:228). The total population must have been over 300,000 with 125-200,000 residing in the urban center, approximately 22,000 in the immediate vincinity (rural Teotihuacan Valley), and perhaps 80,000 in the other Basin of Mexico regions (Texcoco, Ixtapalapa, Chalco-Xochimilco, Azcapotzalco, Tenayuca-Cuautitlan, Zumpango-Pachuca, and Temascalapa).

The demography of the Morelos regions, Tula region, "Teotihuacan Corridor," and "Rio Balsas route" within the "empire" probably accounted for at least an additional 30,000. Millon (1981:228) stated that "a total population of perhaps 300,000-500,000 seems a reasonable approximation," while Sanders, Parsons, and Santley's data (1979:108-129, 183-219, Table 6.18) suggested at least 234,300 and probably 250,000 in the Basin of Mexico alone. Including the Tula region, Sanders estimated a Classic Teotihuacan (Middle Horizon) population of 260,000 (1981:176).

There are many dependent and extraneous variables implicit in calculating potential demography, especially because of the lack of actual site surveys in the western Basin of Mexico and the nature of Classic Teotihuacan settlements in the Toluca Valley to the west. Whether the "Rio Balsas route" had Classic settlements similar to those of the "Teotihuacan Corridor" is not know. Nonetheless, the estimate of over 300,000 within the probably "empire" is valid in terms of the "game" of archaeological demography. However, I believe that Teotihuacan's influence in West Mexico, the Gulf Coast, the Valley of Oaxaca, Guatemalan Highlands, and Yucatan Lowlands may have ultimatey touched the lives of over one million Mesoamericans, especially through the control of the production and distribution of obsidian and spread of ideology and iconography (Kolb 1984a:209-210, 1985,1986:173-175, 196-197).

The "primate" center was the "hub" of the polity, with foodstuffs, raw materials, and foreign products flowing into the city in great quantities. Critical resources (obsidian, lumber, lime, minerals, pigments, salt, etc.) and highly valued raw materials (marine shell, mica, magnetite, iron pyrite, turquoise, flint, chert, jadite, onyx, amber, cinnabar, amate (bark paper), copal (copalli incense), huele (rubber), ocotl (pitch pine), gums and resins, medicinal herbs, dye-stuffs, tropical bird feathers (quetzal, macaw, parrot, etc.), animal skins (jaguar, ocelot, etc.), and eagle feathers, etc.) were imported (Kolb 1986:174, 196). In addition, "exotic" foodstuffs (cacao "beans," fish, waterfowl, non-local game, octli (pulque, etc.) and later, as carrying capacity was exceeded, basic foodstuffs (maize, legumes, vegetables, and fruits) were also imported.

Various "foreign" finished products, especially "Thin Orange," "Fine Orange," Lowland Maya ("Maya style," Tzakol, Mojara Orange polychrome), Gulf Coast (Huastec III), and West Mexican (Chalchihuites) ceramics were brought into the urban center and some rural Teotihuacan Valley sites (Kolb 1964, 1965a, 1965b, 1972a, 1979a:218-227). Barbour (1976:138, 179, 183, 216) reported foreign figurines from Michoacan, Veracruz, the "Gulf Coast," and Valley of Oaxaca. He has, however, postulated that Site 3:S1E3 was a workshop for the production of foreign figurines (1976:216). Borhegyi (1971:85), considering Teotihuacan's monopoly of key products, believed that hallucinogenic mushrooms and peyote were also imported.

Classic Teotihuacan's direct involvement in obsidian exploitation, centralization of workshop production, and subsequent distribution were significant to the commercial life of the "state" and "empire." However, the political and economic lifeways of the Teotihuacanos was inextricably allied with religious ideology. The Teotihuacanos may have been the first Mesoamericans to "systematize religious belief" (Pasztory 1978:9). Whether the belief and its iconography was associated with an agricultural fertility cult that became a "state" cult or religion is open to question (Millon 1981:230). Millon considered that the shrines and temples along the Miccaotli became the center for large-scale pilgrimages so that the "state" religion would "... have embodied a system of belief and ritual that had meaning and appeal not only to regional administrators and foreign rulers ... but to people in a wider range of statuses in the societies with which the Teotihuacanos were in contact" (1981:231). The sytematized "state" religion and the "theocracy" controlled the social and economic lives of the inhabitants of the polity and "empire." The long-distance import and export of raw materials and finished products, including marine shell, was controlled to some degree by the state. The exchange system can be characterized as asymmetrical symbiosis, at least in terms of raw materials, foodstuffs and other tangibles; much was imported into the Supra-Regional Center, but apparently little was exported, but included obsidian and ideology.

CHAPTER TWO:

MAJOR MOLLUSKS

Classifications and Marine Faunal Provinces

Modern zoologists and malacologists (Keen 1971:13-19) have divided the Phylum Mollusca into seven Classes: 1) Gastropoda ("Stomach-footed" [snails], containing three Subclasses and fifteen Orders); 2) Pelecypoda ("hatchet-footed" [bivalves or clams], containing five Subclasses and seven Orders); 3) Cephalopoda ("head-footed" [octopi and squids], containing one Subclass and one Order); 4) Aplacophora ("without plates," [solenogastres or worm-like mollusks lacking shells], containing one Order); 5) Monoplacophora ("bearing one plate" [gastropod-like mollusks], containing one Order); 6) Polyplacophora ("bearing many plates [chitons or sea cradles], containing two Orders); and 7) Scaphopoda ("boat-footed" [tusk or tooth shell], containing two Families). The last four classes are rare, while Cephalopoda are difficult to identify outside of marine contexts. Gastropoda and Pelecypoda are represented in marine environments but have also moved out of the sea into freshwater conditions, but only some gastropods have evolved into terrestrial habitats.

The Gastropoda (snails) evolved during the early Cambrian about 450 million years ago, whereas the Pelecypoda (bivalves or clams) are recorded earliest at about 100 million years ago. Both Gastropoda and Pelecypoda as fossils and live specimens were collected by human beings at least as early as the Upper Paleolithic stage in the Old World and as early as the Lithic IV stage in the New World. "Common" and "rare" or unusual mollusk shells, including Spondylus gaederopodus Linnaeus, 1758, were carried by humans from the Aegean Sea into northern Europe in early Neolithic times (Shackleton and Renfrew 1970), while other Spondylidae were transported from the Gulf and Pacific coasts into central Mesoamerica during the Formative or Preclassic. The collection of living mollusks and/or the shells of deceased snails and clams were transported from their marine habitats into Central Mexican sites during the Preclassic, a practice which continued and apparently intensified during the Classic Period.

The west coast of present-day Mexico, including the Pacific Ocean and Gulf of California coasts of Baja California, and the coastline south through Guatemala and the Central American nations to Panama is considered as one marine ecological zone. This is the Panamanian Marine Faunal Province (P.M.F.P.), also called the Panamic or Panamic-Pacific province, which continues southward into Peru (See Figure 7). The Gulf Coast of the United States and Mexico, the Gulf of Campeche, the Campeche Bank, and Carribean Coast through the Central American nations to Panama is the Caribbean Marine Faunal Province (C.M.F.P.). Both faunal provinces are sources for all seven classes of Mollusca, but especially the Gastropoda and Pelecypoda. Marine mollusks frequently live in restricted ecological niches, particularly defined in terms of ocean currents, water temperatures, and habitat depth ranges. Therefore, in some cases, more specific locations within the marine faunal provinces can be determined. In both Mesoamerica (Millon 1981:227) and the Andean Region (Paulsen 1974) Prehispanic peoples especially sought three genera, Spondylus Linnaeus, 1758, a Pelecypoda; and Strombus Linnaeus, 1758 and Fasciolaria Lamarck, 1799, both Gastropoda. These genera are represented in both the Panamanian and Caribbean Marine Faunal Provinces.

Spondylidae

Six species or subspecies of Spondylus Linnaeus, 1758 were found in the Panamanian Marine Faunal Province (Morris 1966:128-129; Keen 1971:96-98). These six are members of Class Pelecypoda, Subclass Pteriomorphia ("wing-shaped"), Order Pterioida ("winged shells"), Superfamily Pectinacea, Family Spondylidae. The largest in size and widest ranging of the six is Spondylus calcifer Carpenter, 1857 (Synonyms: S. limbatus Sowerby of Reeve 1856; S. radula Reeve, 1856; S. smithi Fulton, 1915). Spondylus calcifer, known as the "Pacific Thorny Oyster," in its adult stage is 15.0 cm across and weighs 1.5 kg or more, and characteristically has numerous, evenly distributed spines. A wide purplish-red band marks the interior margin of most specimens. This species ranges from the Gulf of California southward to Ecuador. "The name 'calcifer' (lime-bearer) refers to the extensive use that was made by Spanish settlers of Central America, who used the lime of these shells as a source of cement" (Keen 1971:96).

Spondylus princeps Broderip, 1833 (Synonyms: S. dubius and S. leucacantha Broderip, 1833), also called the "thorny oyster," in its adult stage is 10.0-15.0 cm across and weighs less than 1.0 kg. Characteristically the hinge ridges are so interlocked that the valves cannot be separated without breaking the teeth. It ranges in the southern Panamanian province, especially

FIGURE 7: MARINE FAUNAL PROVINCES OF THE MESOAMERICAN
            CULTURE AREA

(after Abbott 1962:18-19, 1968:35)

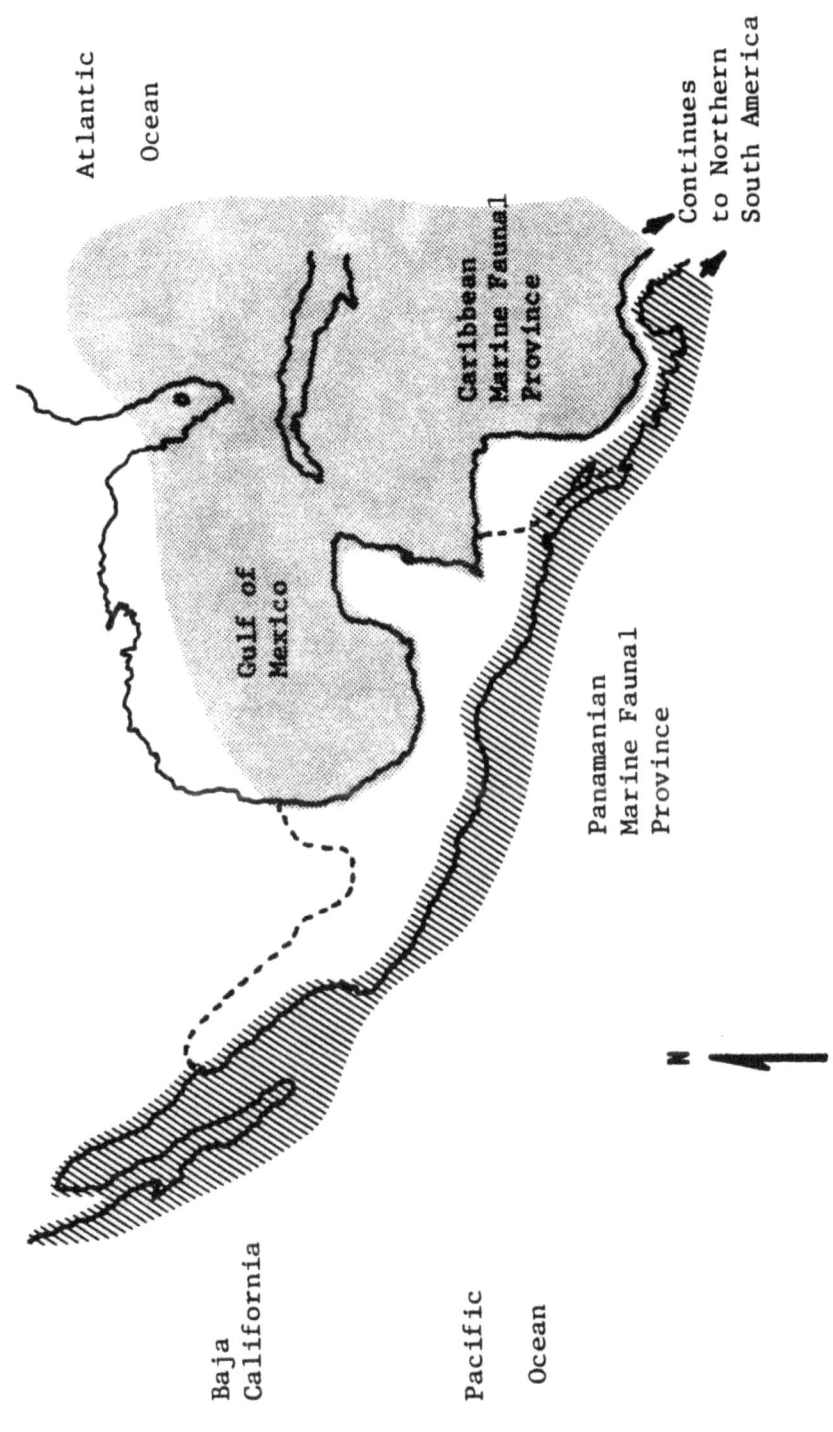

off Ecuador, and may be a subspecies of the more common S. calcifer. Spondylus princeps princeps Broderip, 1833, in its adult stage is 13.0 cm across and has a uniform coral red color, but is restricted to the zone between Panama and northwestern Peru. Spondylus princeps unicolor Sowerby, 1847, is the "Common Thorny Oyster" of the Gulf of California. Its exterior color varies from white through orange or pink to red, but the colored inner band is narrow to absent. This subspecies is smaller than the other Spondylidae, has a different arrangement of spines, and is found no further south than the coast of Jalisco. Spondylus ursipes Berry, 1959 is similar to S. calcifer, averages 10.0 cm in width, and has a narrow orange inner band. This species is restricted to Angel de la Guarda Island, Gulf of California. The last Spondylidae, S. gloriosus Dall, Bartsch and Rehder, 1938, is a Hawaiian Islands species found only as far east as Clipperton Island.

Spondylus americanus Hermann, 1781, the "Atlantic Thorny Oyster," was the only Spondylidae represented in the Caribbean province (Morris 1951:22-23, Warmke and Abbott 1962:170-171). The species has a heavy shell and is 7.0-10.0 cm in adult width with colors ranging from all white, or white with yellow or orange-red umbones, to all orange or all red. The numbers and lengths of the spoines are variable. This species ranges from coastal Florida through the West Indies, the Gulf Coast, and the Caribbean coast of Central America.

Of the six Panamanian and one Caribbean species, Spondylus calcifer and Spondylus princeps from the former province and their Atlantic cousin, S. americanus, were the commonest and most exploited of these pelecypods. The Caribbean species was smaller but more colorful than its Panamanian counterpart. Spondylus calcifer was particularly distinguished by its wide purplish-red interior band, a characteristic lacking in all other Spondylidae species. All species from both provinces live in warm seas attached by their right valves to rock, coral, or other hard objects at depths ranging from 7.0-60.0 m. Both S. calcifer and S. princeps in their adult stages frequent depths of 10.0-25.0 m and often are riddled by the burrows of sponges, marine worms, and small boring clams. Frequently Spondylus becomes "home" for the Family Vermetidae, Genera Vermetus Daudin, 1800 or Petaloconchus H. C. Lea, 1843, the "Black Worm Shell." This marine "worm" (actually a snail) often occurs in colonies attached to rocks or shells, and is notably associated with Spondylus calcifer (Abbott 1962:38, 1968:84).

Contemporary, non-SCUBA-equipped "shell divers" venture to depths of fifteen to twenty fathoms (25.5+ m) searching for pearl oysters in the Gulf of California (Hubbs and Roden 1964:178). However, Spondylus shells are commonly found on beaches after the specimens have died and are cast ashore. Although these have lost their prized "feathery" fronds, they still represent a valuable resource of raw material for processing into artifacts, particularly jewelry. Gathering these shells would have been a relatively simple task, and distribution to inland areas of the Meseta Central and/or Guatemalan Highlands and southern Yucatecan Lowlands would have been easily accomplished.

Spondylus, called teotlchipuli in Nahua, meaning "divine conch" (Seler 1963(2):71), evoked the planet Venus, "Señor de la Aurora," according to Sejourne 1966c:161). Because of this apparently sacred association, Spondylus shells were associated with ofrendas (offerings), used as grave goods in human interments, and occurred in caches with and without other material culture.

Strombidae

The genus Strombus is represented by four species in the Panamanian Marine Faunal Province (Morris 1966:165-166, Keen 1971:420-421). These species are members of Class Gastropoda, Subclass Prosobranchia ("gills forward"), Order Mesogastropoda ("middle stomach footed"), Super-family Strombacea, Family Strombidae. The shells are thick and solid, with a narrow aperture and relatively large body whorl, and the mollusks are herbivorous, grazing on bottom deposits and fine algae. The Subgenus Strombus includes only Strombus (Strombus), gracilior Sowerby, 1825, which is yellowish-brown in color, has a light central band, and white aperture edged in orange-brown. These small strombs average 7.5 cm in length, range from the Gulf of California to Peru, and resemble a Caribbean species S. pugilis Linnaeus, 1758.

The Subgenus Lentigo includes only Strombus (Lentigo) granulatus Swanson, 1822, which typically has brown spots on a whitish or violet colored background. It is a more slender version of S. (S.) gracilior and ranges from the northern end of the Gulf of California to Ecuador. The Subgenus Tricornis contains two species. Strombus (Tricornis) galeatus Swanson, 1823 (Synonyms: ?S. crenatus Sowerby, 1825; S. galea Wood, 1828), is the heaviest and largest Panamanian gastropod. Adult specimens are ivory white, while young shells are variegated with brown on white, banded or blotched in orange-yellow. The adults average 19.0 cm in length and range from the northern end of the Gulf of California to Ecuador. A similar Caribbean species is S. goliath Sowerby, 1842. Strombus (Tricornus) peruvianus Swanson, 1823 adults are 15.0 cm in length and have tan or brown shells. This species ranges from Tres Marias Islands, Mexico, to northern Peru.

The Strombidae are represented in the Caribbean Marine Faunal Province by five species (Morris 1951:164-166, Warmke and Abbott 1962:88-89). Strombus gigas Linnaeus, 1758 (Synonyms: S. samba Clench, 1946; S. gigas verrilli McGinty), is a massive shell 15.0-30.0 cm in length as an adult. The aperture is large and flaring, and the shell brownish-colored with shades of pink, peach, or yellow. Also called the "Queen Conch," this species ranges from Bermuda and the West Indies through the Gulf of Campeche and Caribbean shores of Central America. Strombus pugilis Linnaeus, 1758, the "West Indian Fighting Conch," is a bright, deep orange color and attains an adult length of 7.0-10.0 cm. Its range includes the West Indies, the Gulf Coast and northern Caribbean shores of Central America.

Strombus ranius Gmelin, 1791 (Synonym: S. bituberculatus Lamarck) is the "Hawk-wing Conch" and has a length of 5.0-10.0 cm in the adult. The shell is grayish with brown mottling, and the species is common to the West Indies, eastern shores of the Yucatan Peninsula and the Caribbean coast of Central America as far south as the Orinoco River in Venezuela. Strombus costatus Gmelin, 1791 (Synonym: S. spectabilis Verrill, 1950), the "Milk Conch," is a very heavy shell 10.0-18.0 cm in adult length and has a characteristic yellowish white color. This species ranges from southeastern Florida through the West Indies. The fifth species Strombus gallus Linnaeus, 1758, the "Rooster-tail Conch," is a rare species 10.0-16.0 cm in adult length and grayish brown in color. It is rare in southeast Florida, but more common in the West Indies.

Of the four Panamanian and five Caribbean species, Strombus (Tricornis) galeatus from the former province and Strombus gigas from the latter were the largest conchs and those most sought by Prehispanic Mesoamerican peoples (Pasztory 1978:13). The smaller Strombus pugilis from the Caribbean province was also collected. The Strombidae live in shallow water from tidal areas to depths of ca 10.0 m, except for S. gracilior which lives in depths from 10.0-45.0 m.

Fasciolariidae

Three species of Fasciolaria were represented in the Panamanian Marine Faunal Province (Morris 1966:188, Keen 1971:611), and are members of Class Gastropoda, Subclass Neogastropoda ("new stomach-footed"), Superfamily Buccinacea, Family Fasciolariidae, Subfamily Fasciolariinae. The Genus Fasciolaria Lamarck, 1799 contains one Subgenus, Pleuroploca Fischer, 1884. The species include F. (P.) granosa Broderip, 1832, a heavy, dark shell 12.5-17.5 cm in length, which ranges on mud flats from the Gulf of Californis to Peru. A second, larger species is F. (P.) princeps Sowerby, 1825 (Synonym: F. aurantiaca Sowerby, 1828 not Lamarck, 1816). These orange-brown gastropods attain adult lengths of 15.0-22.5 cm with some specimens up to 30.0 cm, and range offshore from the Gulf of California to Peru. The third Panamanian species, F. (P.) salmo Wood, 1828 (Synonym: F. valenciennesi Kiener, 1840), is colored light orange to yellowish, and attains lengths of 10.0-12.5 cm. This species ranges offshore from Acapulco to Panama.

Two Caribbean species of Fasciolaria occur in the Gulf of Mexico (Morris 1951:207-208; Warmke and Abbott 1962:119). Fasciolaria tulipa Linnaeus, 1758, the "True Tulip," is a shallow water species of variable color (orange, red, or mahogany brown clouded with white, yellow or tan), averaging 7.0-22.0 cm in adult length. F. (Pleuroploca) gigantea Kiener, 1840, the "Florida Horse Conch," is a twin species of the Panamanian F. (P.) princeps. It is brownish orange and has average lengths of 15.0-24.0 cm but some specimens attain lengths of 30.0 cm.

Fasciolaria (Pleuroploca) gigantea was reputed to be the symbol of the Aztec moon deity Tecciztecatl, and the shell itself was called tecciztli (Linne 1942:151). Located on the shore of Lake Texcoco southwest of Teotihuacan, was an Aztec Postclassic village named Tequizistlan, in earlier times called Tecciztlan, the "place where shells abound." Nuttall (1926:58, 61) stated that its name derived from the fact that "there are many shells in the canals of said town." This species of Fasciolaria (F. (Pleuroploca) gigantea Kiener, 1840), the "Florida Horse Conch" has a Caribbean habitat (Warmke and Abbott 1962:119; Keen 1971:611), so that the village name and its association with this specific gastropod is questionable. Tequizistlan may have been the site of freshwater clam collecting (Unio discus Lea, 1838) or involved as a canoe station in the importation of shell from the Caribbean during the Postclassic. The village hieroglyph includes a conch shell representation.

Fasciolaria and Strombus were the raw materials for the manufacture of shell trumpets (Linne 1942:151).

Millon (1981:227), following Starbuck (1975, 1977), noted that two bivalves Spondylus and Pecten, and three conch shells, Fasciolaria (Pleuroploca) gigantea, Strombus (Tricornis) gigas, and Turbinella angulata were especially desired for ritual and ornamental purposes. In summary, Millon stated that "long distance trade in shells must have been a major concern of the state. Some kind of control of the shell trade from both coasts in Central Mexico and far beyond seems probable..." (1981:227). The mural art, sculpture, and minor arts of Teotihuacan are next considered prior to an examination of shell procurement and distribution.

CHAPTER THREE

MARINE MOLLUSKS IN THE ART AND ARTIFACTS AT TEOTIHUACAN

Marine Representations in Mural Art

A number of Panamanian and/or Caribbean marine Gastropoda and Pelecypoda were depicted in wall murals or as bas relief sculpture in the ceremonial complex of the Supra-Regional Center, and in the mural art of some urban Teotihuacan residences. Major syntheses of Teotihuacan mural art have been prepared by Villagra Caleti (1951, 1952, 1954, 1956-56, 1971), Miller (1973, 1978), and Millon (1966a) and are relied upon in the following summary. The precise identification of genera and species of mollusks is difficult because the mural painters either stylized the shells or depicted them as decorated, plumed, painted, cut, or otherwise modified. I have suggested classifications only to the genus level, and, sometimes to that taxon with hesitation. For example, in artistic representations of specimens within the Class Pelecypoda, which includes the genera Pecten Muller, 1776 and Argopecten Monterosato, 1889, it was difficult to determine specific genera let alone the more precise subgenera or species, for a total of seven Panamanian and five Caribbean species (Warmke and Abbott 1962:67-68, Keen 1971:84-88). I have characterized these as Pecten.

In Zone 2 of the ceremonial complex between Platforms 4 and 5 in the Plaza of the Moon Pyramid, the "Talud of the Painted Shells" was found at the top of the staircase (Miller 1973:46, Figures 12-14). The mollusks represented were Strombus and, apparently, Spondylus. In the northeast corner of Portico 1 of the Patio of the Jaguars (Millon's Site 76:N4W1), Murals 1 and 4 there were profiles of jaguars with feathered headdresses. (See Figure 8.) Each held a plumed conch shell in one paw and may be blowing it (Miller 1973:50, Figures 24, 26; Millon 1973:Figure 286). The conch was probably Strombus or less-likely Turbinella. A line of stylized marine bivalves decorated the back and tail of this feline. Mural 2 of Portico 2 in the Patio of the Jaguars depicted bivalve shells with serrated ventral margins (Miller 1973:51, Figure 28), which were probably Pecten rather than Spondylus. Mural 1 of Portico 6 in the Patio had a Tlaloc figure within a cross-section of a Pecten or Spondylus (1973:51, Figure 30), and Mural 4 of Room 7b of the associated Palace of the Jaguars had a mural with bivalves and human feet (1973:52, Figure 33). These

FIGURE 8:   PLAN OF THE PALACE AND PATIO OF THE JAGUARS
(76:N4W1)

(after Miller 1973:49, Plan III)

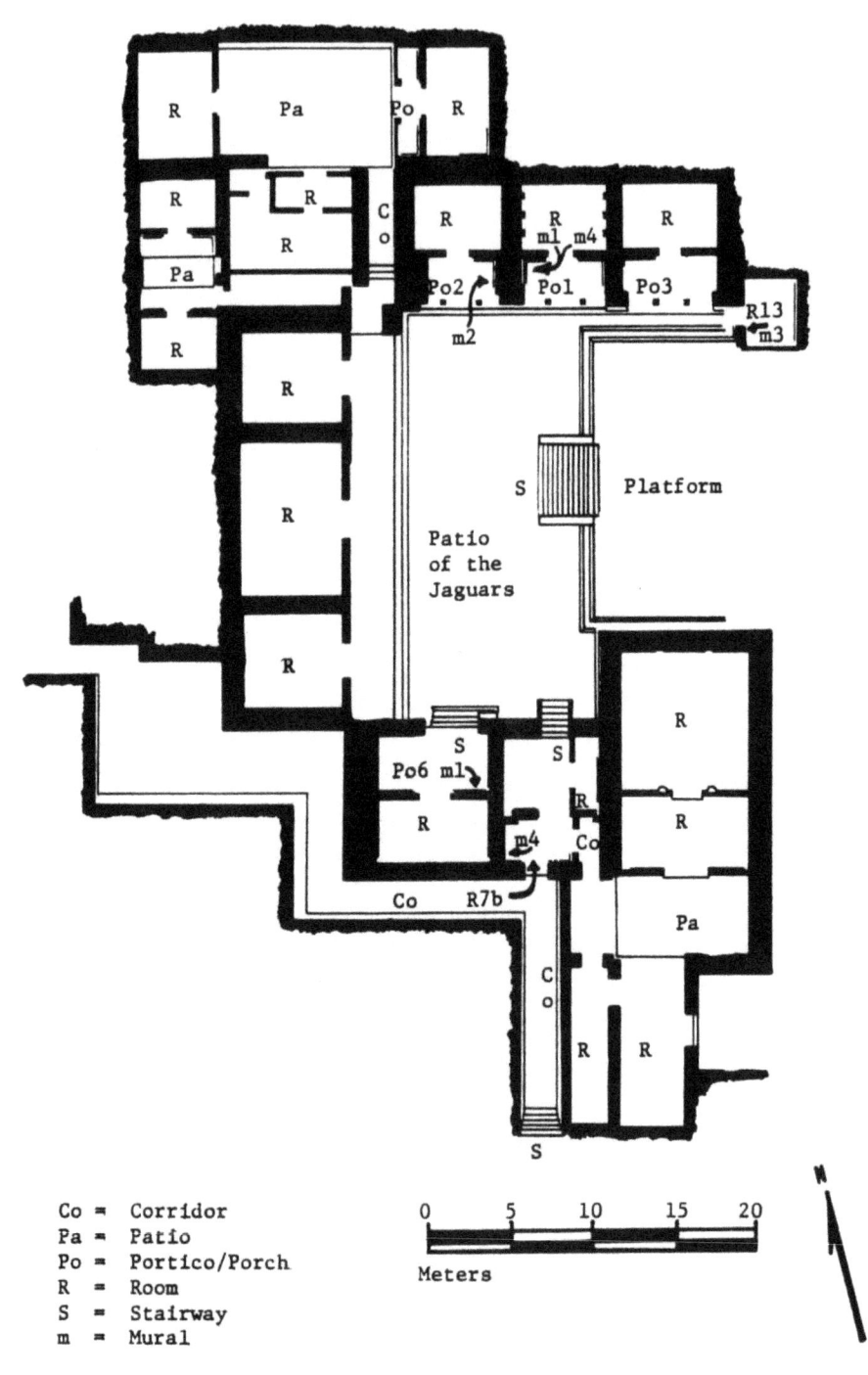

Co = Corridor
Pa = Patio
Po = Portico/Porch
R = Room
S = Stairway
m = Mural

shells were possibly Pecten or Spondylus, and I believe that the depiction of feet may be a symbol indicating long-distance transport or trade with the coast. Mural 3 in Room 13 of Zone 2 had a mural depicting what seems to be an Oliva shell (1973:58, Figures 50-51).

The Temple of Agriculture (Millon's Site 5:N4W1), also located in Zone 2, originally had a painted fresco of ten shells and horizontal turquoise-green "wave" motifs in a corridor called the "Hall of Mollusks" (Herrera 1922:189, Lam. 27, Figures 40a-b, 41a-c; Marquina 1951:90-91, Lam. 22; Villagra Caleti 1971:138-140; Miller 1973:66, Figures 79-80). The mollusks represented include apparent Spondylus associated with the "waves," and conch shells which were of Turbinella or Strombus, as well as other stylized bivalves (Furst 1966:161-163).

Six mollusk representations were found in Zone 5-A structures in the ceremonial complex, Millon's Site 8-9:N4W1. Mural 2 in Portico 3 depicted three genera of shell, Pecten, Spondylus, and Oliva plus a stylized "starfish" (Class Asteroidea) or cross-sectioned conch (Miller 1973:80, Figures 114-115). These genera and "starfish" were also illustrated in Murals 1 and 2 of Portico 13 (1973:80-81, Figures 116-119). Starfish or cross-sectioned conch shells were also found in Mural 3 of Room 13 (1973:82, Figure 121), while Mural 1 of Room 18 had an enormous scallop-like shell and bivalves (1973:84, Figures 125-128). The shell representations included Pecten, Spondylus and Oliva. Lastly, Mural 1 of Portico 22 in Zone 5-A had a cross-sectioned conch or starfish (1973:87, Figure 137).

The Painted Patio at the urban residence of Atetelco (Armillas 1950:56-57; Marquina 1951:94; Villagra Caleti 1951:153-162, 1952:67-74, 1954:69-78, 1956-1957:1-3, 1971:143-144) contained a "miniature temple," actually a platform altar with talud and tablero construction. Atetelco was designated Site 1:N2W3 by Millon (1973). (See Figure 9.) The two reconstructed north side entablature murals depicted twelve distinct shells: seven bivalve Pelecypoda, and three conch and two olive shells, all Gastropoda (Miller 1973:164-165, Figures 345, 347). Among those represented were Strombus, Turbinella(?), Spondylus, Pecten and Oliva. A mural found in a private collection, painted in what Miller considers the "Atetelco style," had a polychrome animal with a speech scroll and had a starfish or cross-sectioned conch shell depicted on its body (Sejourne 1966b: 152, 154, Figure 80; Miller 1973:170, Figure 367).

FIGURE 9: PLAN OF THE ATETELCO SITE (1:N2W3)

(after Miller 1973:158, Plan XIV-A and 163, Plan XIV-B; Millon 1973: Map 1)

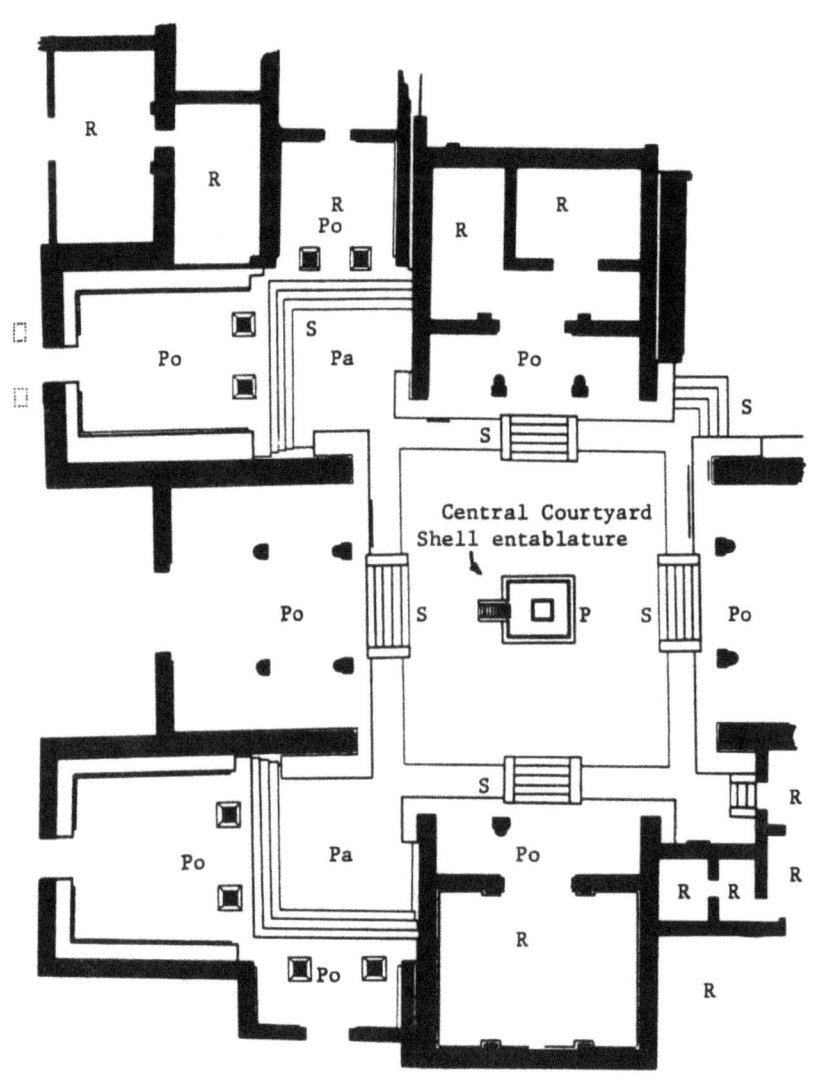

CC = Central Courtyard/Principal Patio
P = Platform Altar
Pa = Patio
Po = Porch
R = Room
S = Stairway

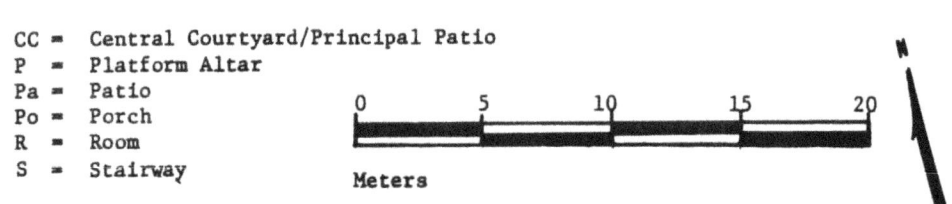

A mural painting at Teopanacazco, Millon's Site 1:S2E2, had priest-like human figures with speech scrolls (Seler 1915:414-417, Abb. 8, Villagra Caleti 1971:138-139). The garments of these figures and the frieze border contained depictions of Pecten, probably Spondylus, Oliva and other stylized bivalves.

The Tetitla urban residence, 1:N2W2, (Sejourne 1966:2-3) had especially diverse marine representations. (See Figure 10.) Mural 4 on the rear wall of Room 7 illustrated a human figure whose torso has been replaced by seven bivalve shells (Miller 1973:134, Figures 268-270). The mollusks were either Pecten or Spondylus. Mural 5 on the west wall of Room 7 depicted two elderly, bearded human figures ("old men") in profile emerging from enormous bivalves (1973:134-136, Figures 271-273). Muller noted that the sashes on these figures and parts of the shells were painted in an unusual red-yellow color rarely seen in Teotihuacan murals, but which was a color common to Tepeu Maya ceramics (1978:68-69). Mural 4 of Tetitla Room 17 (Sejourne 1966:269, Figure 155; Miller 1973:144, Figures 296-297) showed the arms, legs, and head of a "doglike creature" emerging from a decorated bivalve, probably a Pecten or Spondylus. Mural 1 of Room 18 depicted decorated bivalves (Pecten or Spondylus) alternating with speech scrolls (Miller 1973:129, Figure 259), and Mural 1 on the rear wall of Room 22 had two confronting profile birds (eagles or owls) above Strombus conch shells (Miller 1973:127, Figures 250-252).

Mural 5 of Tetitla Corridor 12a had an offering scene including a bivalve (Pecten or Spondylus), green disks (greenstone or jadeite?) and Tlaloc fangs (Miller 1973:155, Figures 326-237). In Patio 5, Mural 1 there were two feathered or plumed conchs (Strombus) confronting a Tlaloc image, while associated Mural 2 had a profiled conch shell (Sejourne 1966b:250-251, 254, 256-257, Figures 137, 142; Miller 1973:132-133, Figures 264-267). A non-provenience mural fragment from Tetitla illustrated two bivalves (Pecten) (1973:157, Figure 331).

Two porticos at Tetitla also had marine representations. Mural 1 from Portico 24 had two "head-on" views or cross sections of conch shells (1973:128, Figures 253-254), which were probably Strombus gigas (Miller 1978:68-69), a Caribbean species. Murals 3 and 4 on Portico 26 each illustrated a human "shell diver" in the process of collecting bivalves which he is placing in a net hung around his neck (Sejourne 1966b:60-61, 302-303, Figures 17, 178; Miller 1973:136, Figures 274-277, 1978:68-69; L. Parsons (1978:30, 32). In the Gulf of California, divers recovered pearls from depths as great as 15-20 fathoms (Hubbs and Roden

FIGURE 10:   PLAN OF THE TETITLA SITE (1:N2W2)

(after Sejourne 1966b:2-3, Fig. 1, and 8-9, Fig. 2;
 Sejourne 1969:92-93; Miller 1973:119, Plan XIII; Millon
 1973, Map 1)

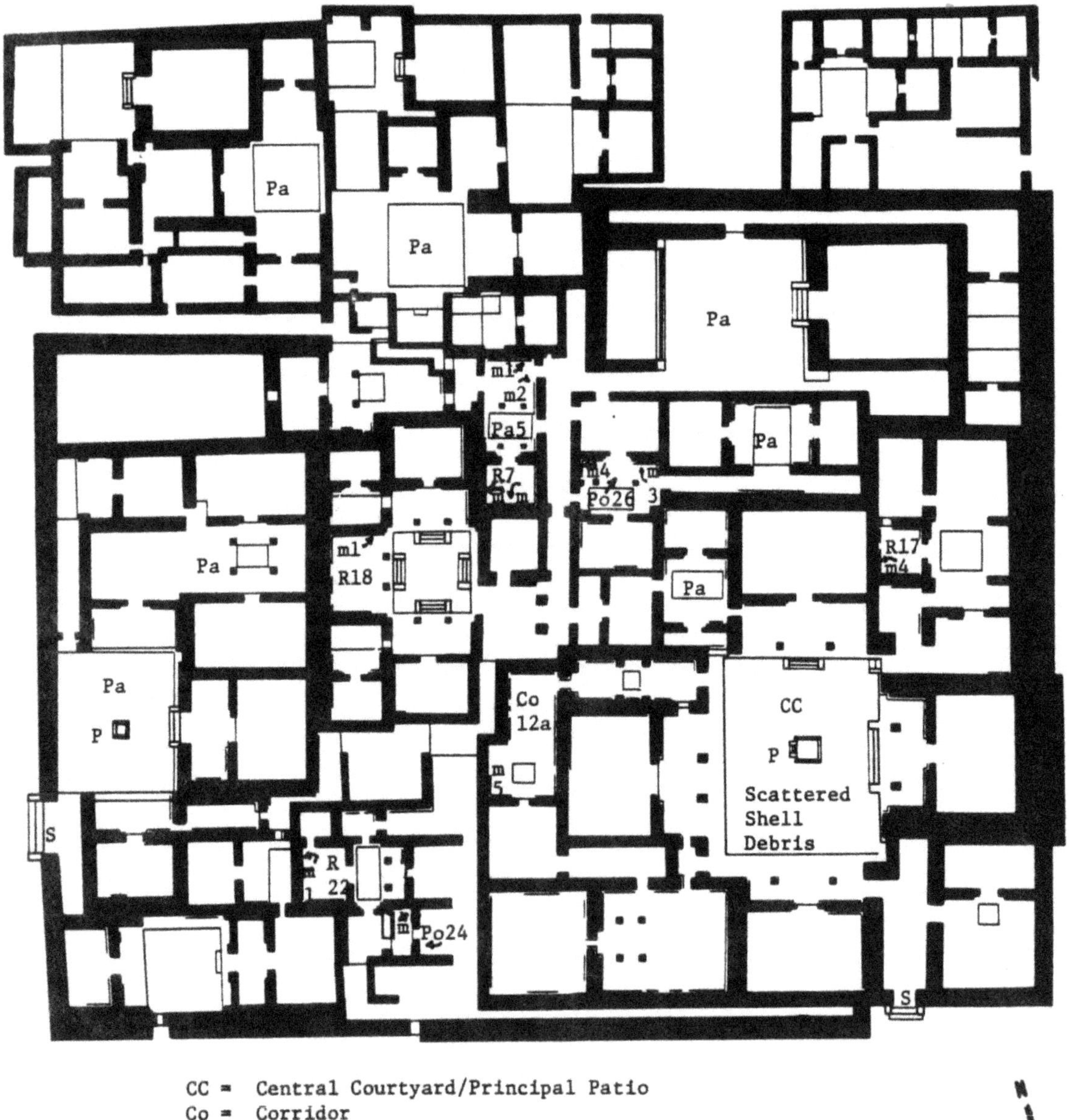

CC = Central Courtyard/Principal Patio
Co = Corridor
P = Platform Altar
Pa = Patio
Po = Portico/Porch
R = Room
S = Stairway
m = Mural

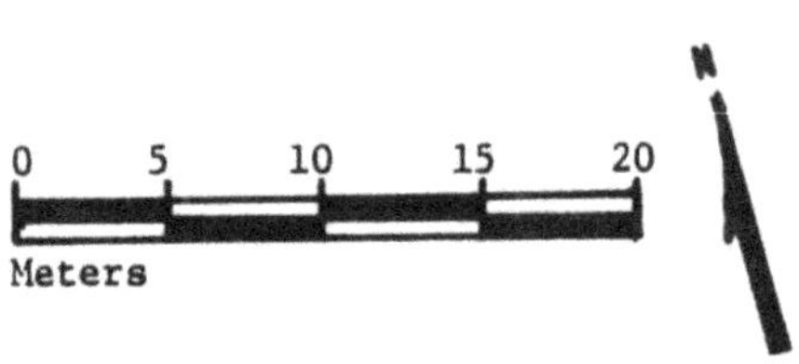

1964:178), so that "shell diving" is not an unusual Mesoamerican activity, even in deep water. Miller stated "that the Teotihuacan painters could show a marine-shell diving scene on a wall on the Mexican plateau 250 miles from the nearest sea is a testament to interconnections between Teotihuacan and the sea coasts" (1973:136). He argued, that, on the basis of the unusual red-yellow pigment in Mural 5 of Room 7 and a "Maya style" seated figure in Room 27, there was at least Lowland Maya influence in the Teotihuacan Tetitla murals (1973:68-69). Two figures in Maya costume were depicted in a reputed Tetitla mural now in the Saenz Mural Collection.

In the Tepantitla residence, 1:N4E2, Mural 6 of Patio 2 had a "starfish" and two types of shell, probably, Pecten and Spondylus (Armillas 1950:55; Marquina 1951:94; Miller 1973:98, Figures 166-167). (See Figure 11.) Other representations were found at the Yayahuala apartment complex, 1:N3W2, (Sejourne 1966b:26-27). (See Figure 12.) Mural 1 of Portico 1 had illustrations of a conch shell in cross-section and other shells similar to those depicted in Zacuala Patio murals, 2:N2W2, (Sejourne 1959:20, 22-23, 28, Figures 2, 4, 10; Miller 1973:107, Figure 197). (See Figure 13.) Mural 4 in Corridor 2 of the Zacuala Patios (Armillas 1950:56; Sejourne 1959, 1966b:18-19) showed a starfish or conch shell in cross-section, as did a mural in Room 2 (Miller 1973:116-118, Figures 220-221, 225-226). In the adjacent Zacuala Palace, 3:N2W2, dated to the Early and initial Late Xolalpan phases (Millon 1966a:11), Mural 2 of Corridor 1 had a Tlaloc figure with a starfish or cross-sectioned conch in a cartouche (Miller 1973:110, Figures 201-202). (See Figure 14.) Mural 8 in Portico 9 of the Palace depicted two representations of nearly identical Tlalocs with the same cartouches (1973:110, Figure 203). Lastly, the Central Courtyard in the Zacuala Palace had depictions of two types of decorated bivalves apparently Pecten or Spondylus (Sejourne 1959:41, Figure 20:4; Miller 1973:112-113, Figures 209-212). The conchs depicted were Strombus.

Marine Representations in Lithic Sculpture

The Palace of Quetzalpapalotl, 2:N4W1, in Zone 2 of the Supra-Regional Center had sculpted stones found at the base of the staircase which included representations of Pecten (Acosta 1964: Figure 21:4, 33, 35:1-1). (See Figure 15.) The Temple of Plumed Conch Shells (Caracoles Emplumadas), dated to the Early Tlamimilolpa phase, depicted bas reliefs of Strombus gigas Linnaeus, 1758 or Turbinella angulatus Solander (Bernal 1963:20, 22-23;

FIGURE 11: PLAN OF THE TEPANTITLA SITE (1:N4E2)

(after Miller 1973:93, Plan IX: Millon 1973: Map 1)

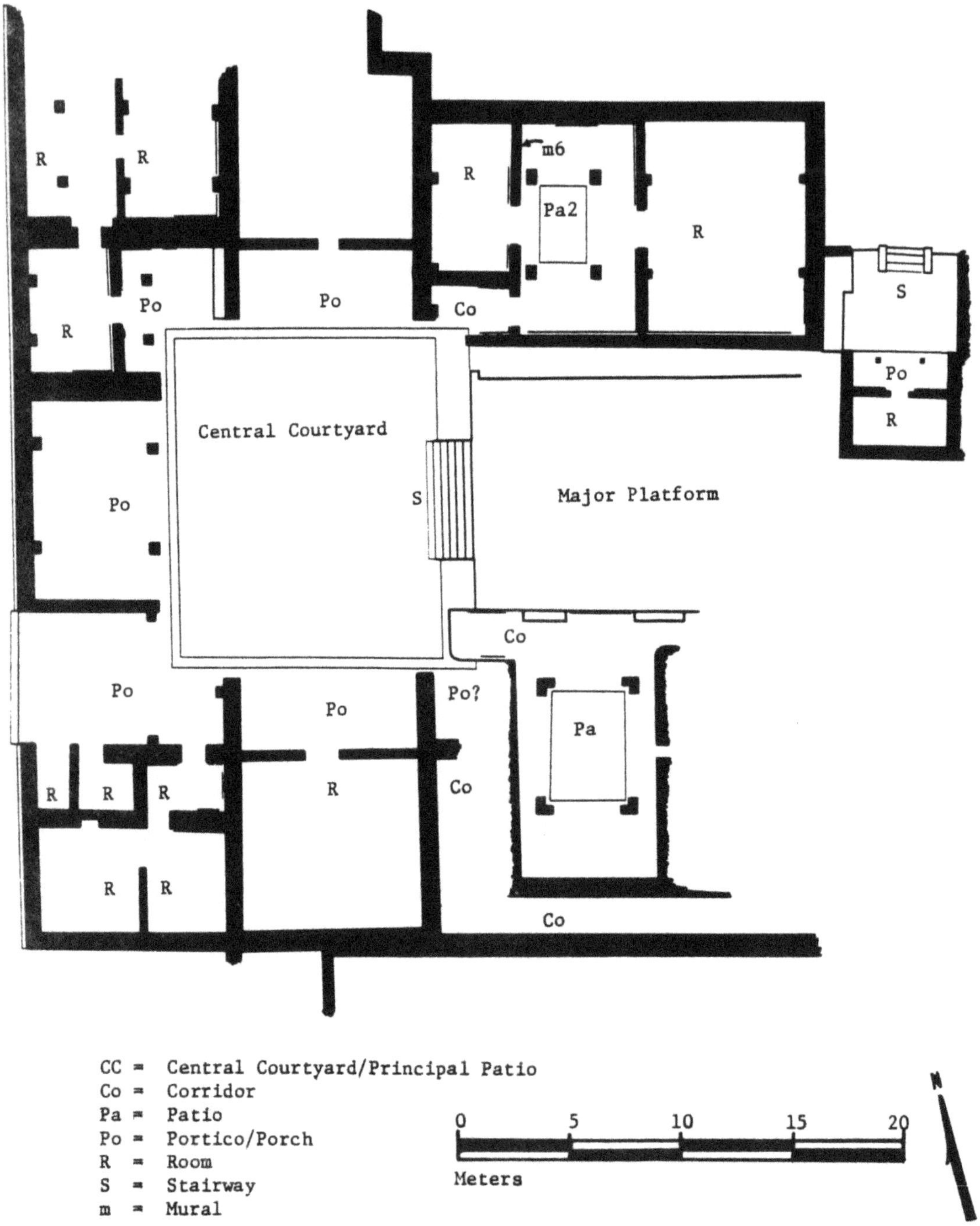

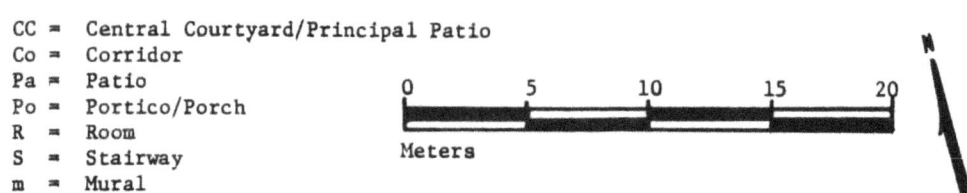

R
R
m6
R
Pa2
R
S
Po
Po
Co
R
Po
R
Central Courtyard
S
Major Platform
Po
Co
Po
Po
Po?
Pa
R
R
R
R
Co
R
R
Co
CC = Central Courtyard/Principal Patio
Co = Corridor
Pa = Patio
Po = Portico/Porch
R = Room
S = Stairway
m = Mural
0
5
10
15
20
Meters
N

FIGURE 12: PLAN OF THE YAYAHUALA SITE (1:N3W2)

(after Sejourne 1969:77; Miller 1973:107, Plan X; Millon 1973:Map 1)

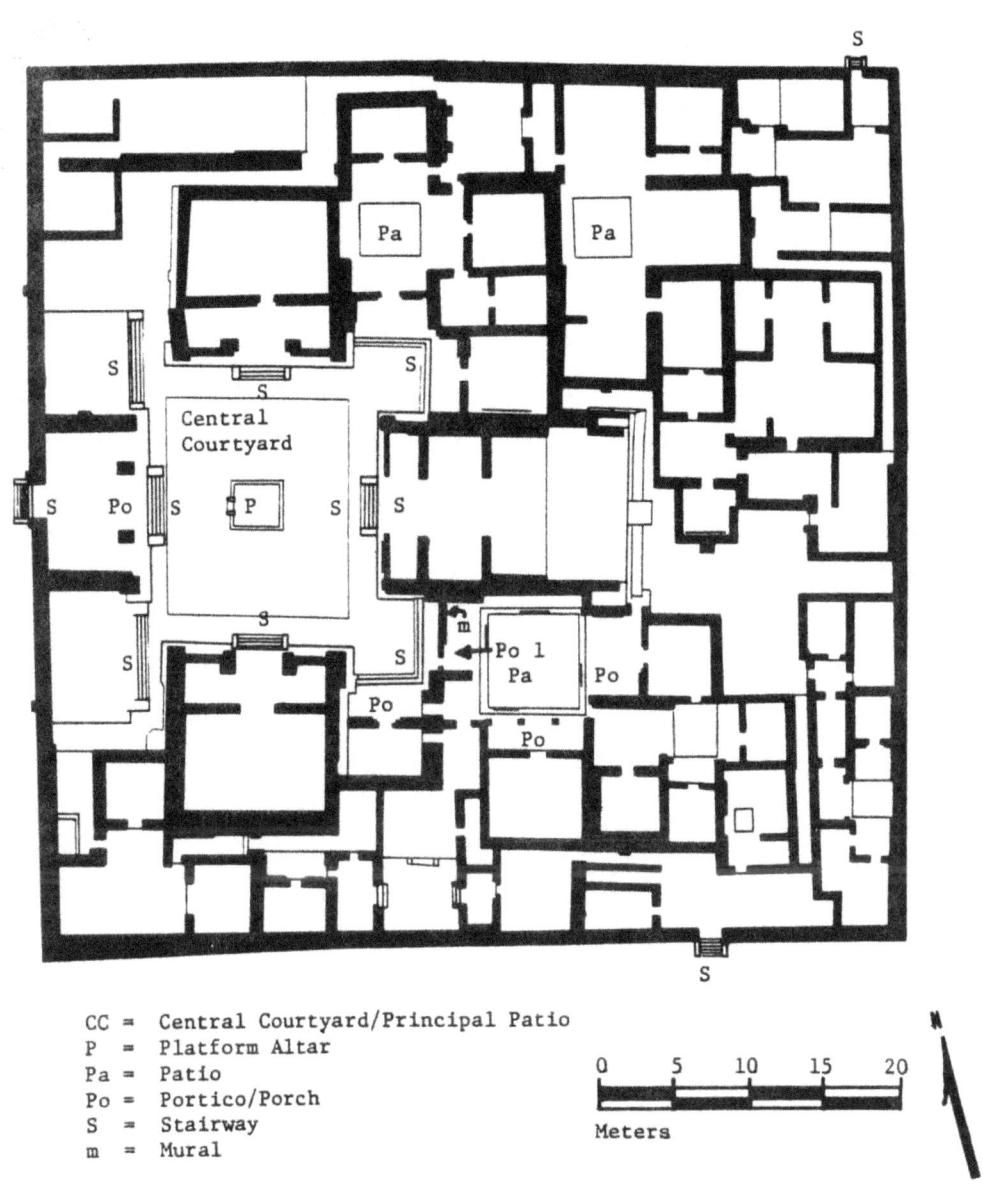

CC = Central Courtyard/Principal Patio
P = Platform Altar
Pa = Patio
Po = Portico/Porch
S = Stairway
m = Mural

FIGURE 13: PLAN OF THE ZACUALA PATIOS SITE (2:N2W2)

(after Sejourne 1959:Fig. 21; Miller 1973:114, Plan XII: Millon 1973:Map 1)

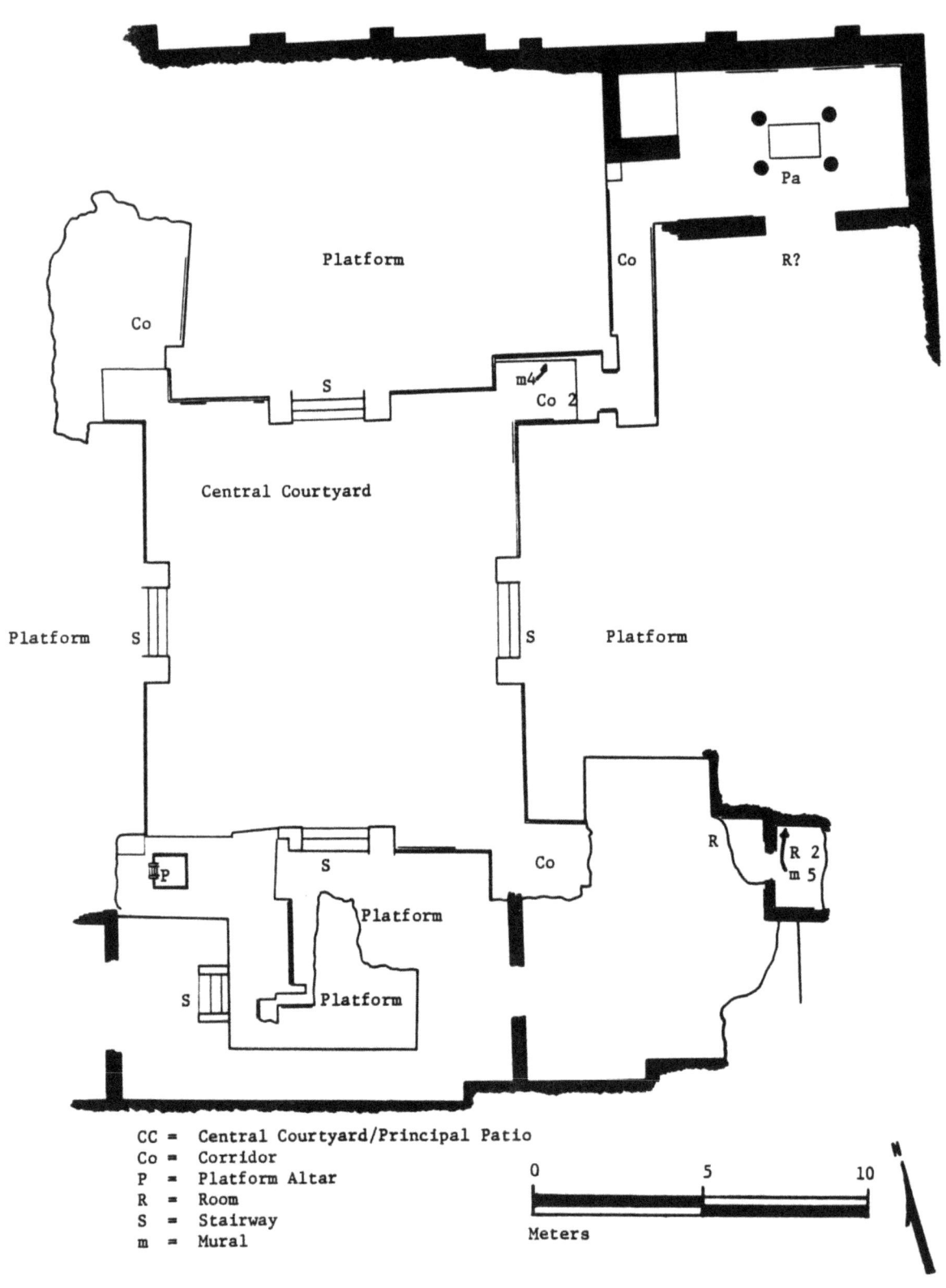

CC = Central Courtyard/Principal Patio
Co = Corridor
P = Platform Altar
R = Room
S = Stairway
m = Mural

FIGURE 14: PLAN OF THE ZACUALA PALACIO SITE (3:N2W2)

(after Sejourne 1959:Figs. 1 and 35, 1966b:18-19, Fig. 3, 1969:76; Miller 1973:108, Plan XI: Millon 1973:Map 1)

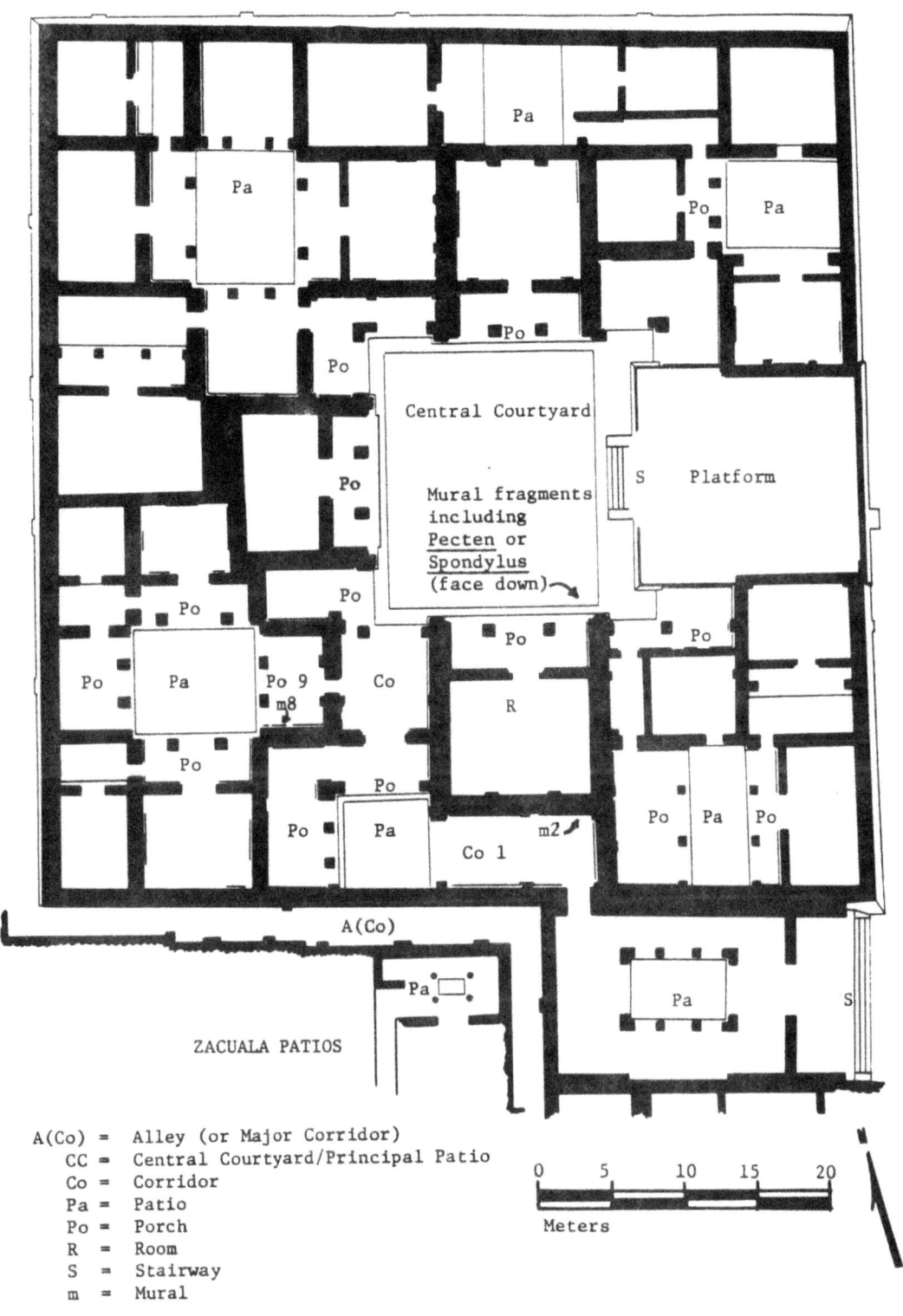

A(Co) = Alley (or Major Corridor)
CC = Central Courtyard/Principal Patio
Co = Corridor
Pa = Patio
Po = Porch
R = Room
S = Stairway
m = Mural

FIGURE 15: PLAN OF THE PALACIO DE QUETZALPAPALOTL
(2:N4W1)

(after Acosta 1964:Plano 10; Miller 1973:42, Plan II;
Millon 1973:Map 1)

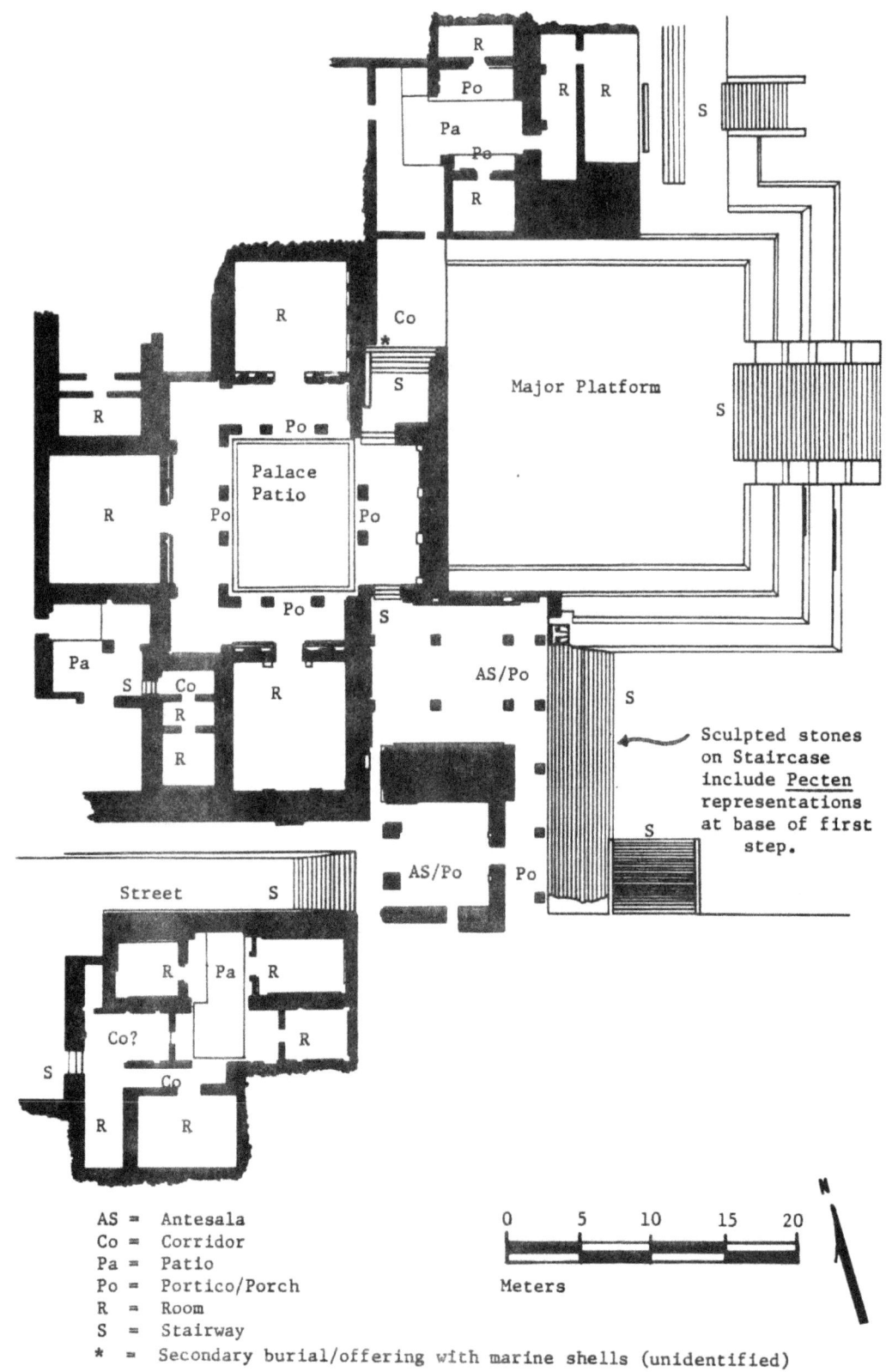

AS = Antesala
Co = Corridor
Pa = Patio
Po = Portico/Porch
R = Room
S = Stairway
* = Secondary burial/offering with marine shells (unidentified)

Volks 1963:94-95; Furst 1966:158-163; Millon 1966a:11; Sejourne 1966b:125, Figure 54). Unprovenienced lithic sculpture studied by Herrera in the local Teotihuacan museum in 1920 also had marine mollusks in bas relief (Herrera 1922:188-190, Lam. 101k, Figures 40-41). Among the species he identified Fasciolaria gigantea (reclassified to Pleuroploca gigantea Kiener, 1840, the "Florida Horse Conch"), which has a range from Florida to North Carolina and was probably misidentified by Herrera (Warmke and Abbott 1962:119, Keen 1971:611). The Panamanian Fasciolaria (Pleuroploca) princeps Sowerby, 1825, a cousin to the Florida genus is a more likely classification. Turbinella scolymus Gmelin and Strombus galeatus (reclassified S. (Tricornis) galeatus Swanson, 1823), both Panamanian species (Keen 1971:421, 620-622) were also identified in Herrera's analysis of the sculpture.

The Temple of Quetzacoatl in the Ciudadela, 1A:N1E1, had an elaborate facade consisting of six tiers of talud and tablero construction and a balustraded staircase facing west. Lithic sculptural depictions of the major deities, Quetzatcoatl and Tlaloc, as well as bas relief representations of marine mollusks occurred on the entablatures (Boekelman 1935:63-65, Figures 2-3; Marquina 1951:81-88; Sejourne 1966b:108-109, Figure 38). The marine bivalves and conch shells retained much of their original paint because the facade was covered during a reconstruction and expansion of the temple pyramid creating the Plataforma Adosada. The bivalves were depicted as half shells with the pink-white interiors exposed, so that the ridges typical of Pecten and spines characteristic of Spondylus exteriors were not visible. Unmodified (ie. non-shell trumpet) white-painted conch were seen in profile. The configurations, colors, and hinge areas suggested that the bivalves were Spondylus rather than Pecten. The conchs were definitely the genus Strombus. Other bivalves, represented in obverse, had ridging characteristic of Pecten (Kolb 1973a).

Marine Mollusks and Ceramic Artifacts

Non-mural representations of marine mollusks occurred as moldmade ceramic adornos associated with elaborate composite censers, appliqued friezes of adornos on pottery vessels, and as incised depictions on ceramics. (See Figures 16 and 17.) Censer adornos were made during the Late Tlamimilolpa through Late Xolalpan phases, and possibly into terminal Classic Metepec phase (Enciso 1947:46; von Winning 1949:141-150, Figures 13-24; Kolb 1965a). Identifiable among the adornos were Pecten

FIGURE 16: EXAMPLES OF MOLD-MADE CERAMIC <u>ADORNOS</u>

(after Sejourne 1966a:Fig. 28; Teotihuacan Mapping
 Project Surface Survey Collections; Teotihuacan Valley
 Project Classic Site Survey Collections)

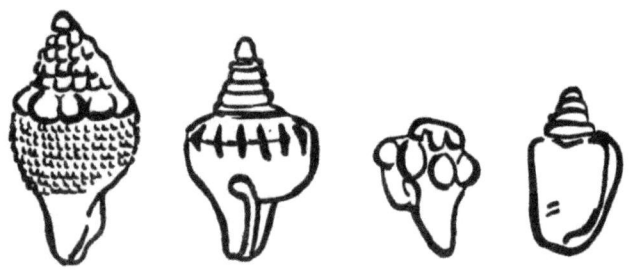

FIGURE 17: EXAMPLE OF A COMPOSITE CENSER WITH ADORNOS
(LATE TLAMIMILOLPA TO EARLY XOLALPAN PHASES)

(after Sejourne 1966a:Fig. 24, Lams. 9, 13; Teotihuacan Mapping Project Surface Survey Collections)

Adornos: Bivalves (12)   Scale 1/4
          Conchs (4)
          Cut Conchs ["Stars"] (2)

(Gamio 1922a:196-200, Lam. 110a, b, d, g, i; Herrera 1922:187-188, Lam. 101a-g; Linne 1934:159; von Winning 1947:333-341, 1949:143-145, Figure 15-18). Herrera suggested specific species: Pecten turgidus Lamarck, 1801 from coastal Veracruz; P. jacobaeus Linnaeus, 1758 found at Mulege, Baja California; P. ventricosus Sowerby, 1842 from La Paz, Baja California, and P. subnodosus Sowerby, 1842 also from Mulege (cf. Warmke and Abbott 1962:167-168, Keen 1971:84-93).

Moldmade ceramic adornos representing olive shells (Oliva sp.) were less common (Gamio 1922a:196-200, Lam. 110n; Herrera 1922:190, Lam. 101l), but actual Oliva specimens were used as personal ornaments and even possible silbatos or whistles (1922:Lam. 101m-n). Molded ceramic adornos of conchs (Strombus and Turbinella), Pecten and cowrie (Cypraea) were also noted (von Winning 1949:141-143, Figures 13-14; Sejourne 1959:126, Figure 106, 1966a:49-50, Figures 27-28, Lam. 13).

Moldmade ceramic adornos depicting stylized and naturalistic shells occurred on basal bands or friezes on Cylindrical Tripod Support Vases in Burnished Black and Brown Wares, Copa Ware, and even on "Thin Orange" Ware cylindrical vases (Kolb 1965a, 1973:332-333, 344-345, 351, 355, 356, 361, 364, 365; Figure 1, Tables VII, IX). The burnished wares, especially vases, jars, and ollas, occasionally had prefired or postfired incised shell designs (Herrera 1922:188, Figures 38-39; von Winning 1949:127-140, Figures 1-9; Cook de Leonard 1971b:216). The depictions of Pecten, Spondylus, Strombus, and Oliva were normally done as simple incision, but also as plano-relief or champleve. Ceramic conch shell trumpets, apparently attempts to replicate Strombus, were made in "Granular Ware" and are recorded at Tetitla (Sejourne 1966b:206, 231-232, Figure 121, Lam. 107; Kolb 1984b:49-50). (See Figure 18.)

Representations of marine mollusks also appeared as decorations on ceramic moldmade figurines (Kolb 1973a, 1973b; Hodik 1974; Sejourne 1966c). Collars and pectoral ornaments on figurines made during the Late Tlamimilolpa through Late Xolalpan phases (ca 400-650 A.D.) occasionally had shells depicted. Bivalves, apparently Pecten, were the most common mollusks (Sejourne 1966c:22, 26, 92, 108, 111, 155, 158-159; Figures 4, 5, 62 left, 76, 79, 111 Row 3:2, 111A Row 3:2-3, 112 Row 3:2), although on Oliva was suspended from a necklace in one specimen (1966c:158, Figure 111A Row 6:1). There were no mollusk decorations appearing in or on headdresses. No ceramic figurine specimens from any excavated or surface collected rural Teotihuacan Valley site had mollusks

FIGURE 18: EXAMPLE OF A GRANULAR WARE CERAMIC TRUMPET
(RECONSTRUCTED) (LATE XOLALPAN PHASE)

(after Sejourne 1966b:231, Fig. 121; Safer and Gill
1982:178; Kolb 1984b: Fig. 1)

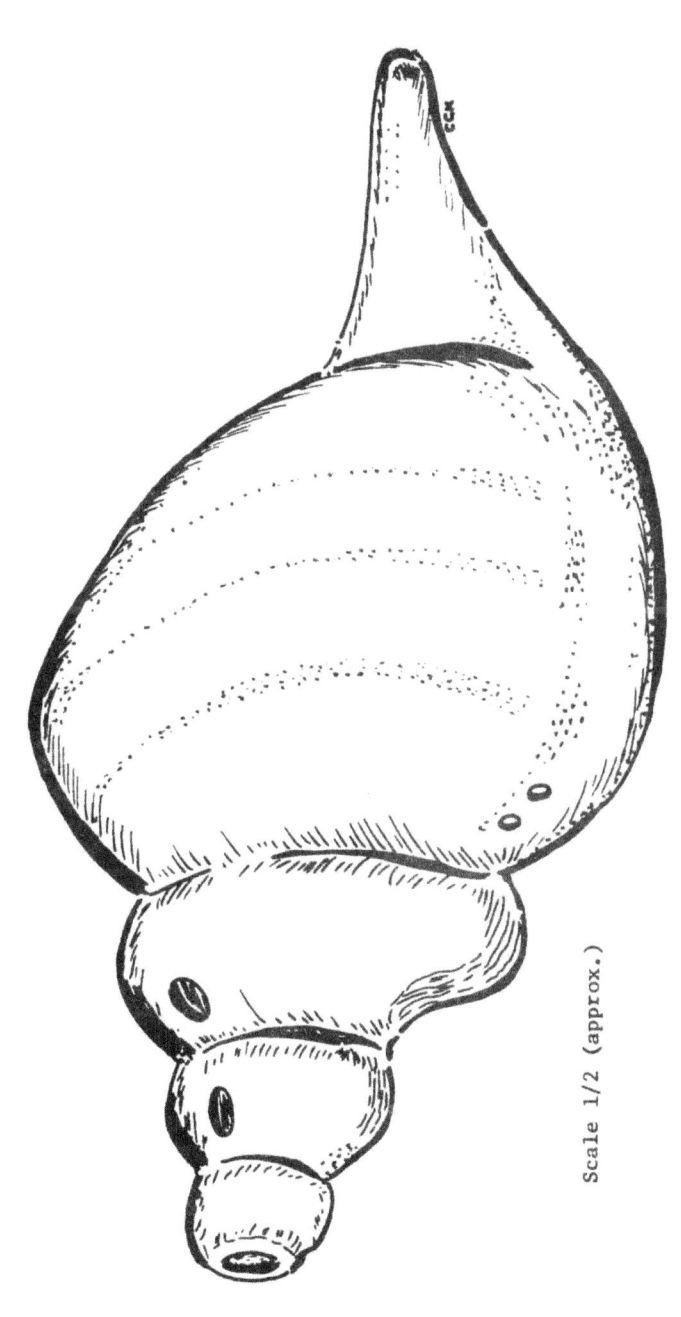

Scale 1/2 (approx.)

associated with headdresses, collars, necklaces, bracelets, etc. (Kolb 1973a, 1973b).

Sejourne's illustration of earspools which have a cross-sectional conch or starfish motif (1966c:85, Figure 56) were, in reality not ear ornaments but clay stamps (1966c:204-211). I am unaware of any mollusk decorations or motifs on earspools, and extremely few depictions on relief stamps (Enciso 1947).

The marine mollusks depicted in frescoed murals, as censer _adornos_, incised decorations, ceramic figurine ornaments, and lithic bas relief sculpture predominantly included Gastropoda and Pelecypoda from either or both marine faunal provinces. Species within the genera Oliva, Pecten, Spondylus, Strombus inhabit both provinces, although numerically more species are associated with the Panamanian. Miller's suggestion of Lowland Maya influence in Tetitla mural art, on the other hand, would indicate a potential Caribbean connection. Given the stylized and decorated mollusks depicted, identification of faunal province was not possible. The mollusks depicted in Teotihuacan art may represent genera from both provinces. The actual identifiable shells in contexts in the Supra-Regional Center argue for both coasts, and for an early and seemingly more intense importation from the Panamanian Marine Faunal Province.

The "Maguey Priest Mural," now located in the Cleveland Art Museum, one of three identical murals, reputedly came to a structure 500 m east of the Pyramid of the Moon (Miller 1973:170, Figure 366). The priest's "speech scroll" contained shells (three bivalves and one gastropod) and flowers. A duplicate, unprovenienced "Rain Priest Fresco" dated ca 300 - 600 A.D., now in the Art Institute of Chicago (Primitive Art Purchase Fund 1962.702), had an identical scroll, and the priest carried an incense bag in his left hand, while water and flowers issued from the right. These were probably fertility murals, and both were rendered in the dark red, red, and pink colors reminiscent of the Tetitla and Tepantitla murals.

A review of the architectural plans indicated that murals with shell and maritime motifs ("shell diver," etc.) tend to be located in interior residential units designated as Porches (n = 11), with others in Rooms (n = 5), and the remainder in a Patio and a Central Courtyard (one each). Therefore, "shell" murals were not placed so as to be seen by the "general public," but were located to be seen by the inhabitants of the residential complex.

CHAPTER FOUR

SPECIFIC MOLLUSK SPECIMENS AT TEOTIHUACAN

Burials and Caches at Urban Residences

Various authors concurred that "huge quantities" of marine shell were imported from their coastal habitats into Teotihuacan during the Classic period (Gamio 1922a:187-190; Linne 1934:158-159, 1942:150-153; Sejourne 1959:65, 68-69, 1966c:156-165; Piña Chan 1963:52; Borhegyi 1971:85, 104; Cook de Leonard 1971:216; Noguera 1971:259; Kolb 1973a, 1979:223-224, 233, 240; Millon 1973:23, 45, 1981:227; Drennan 1984a:38). Marine shell was an important raw material for the manufacture of ornaments, jewelry, mosaics, and other art work, but also was used in caches and as mortuary offerings.

Linne's Excavations

The excavation of the Xolalpan apartment complex by Linne in 1932 produced nine genera of marine mollusks which, with one exception, came from the Pacific Marine Faunal Province (Linne 1934:158-159, Keen 1960, 1971). Millon (1973) designated Xolalpan as Site 2A:N4E2. No specimens were found in the seven Classic Teotihuacan graves, and only a few shells were found below the floors. (See Figure 19.) The only Caribbean species represented was Turbinella scolymus Gmelin (Warmke and Abbott 1962:148-150). The Pacific Coast specimens included four Gastropoda: Fasciolaria princeps Sowerby, 1825; Latirus ceratus Wood, 1828; Melongena patula Broderip and Sowerby, 1829; and Natica (Stigmaulaux) broderipiana Recluz, 1844 (Keen 1971:473, 603, 611). The four Pelecypoda were: Meleagrina margaritacea Lamarck, 1819; Pecten subnodosus Sowerby, 1835; Spondylus pictorum Chemnitz, 1856 = S. princeps Broderip, 1833; and Spondylus princeps Broderip, 1833 (Keen 1971:77, 85, 96). No numerical quantities of Mollusca were reported, but he stated that the "resultant collection is only of slender proportions" (Linne 1934:158). No Spondylus calcifer Carpenter, 1857 were found.

By comparison with the Xolalpan site, Linne's excavations at the Tlamimilolpa residential complex during the 1934-1935 field season produced "quantities" of marine shell in various contexts including graves and caches (Linne 1942:132, 133, 135, 140, 141, 142-143,

FIGURE 19: PLAN OF THE XOLALPAN SITE (2A:N4E2)
(after Linne 1934:42, Fig. 9; Millon 1973:Map 1)

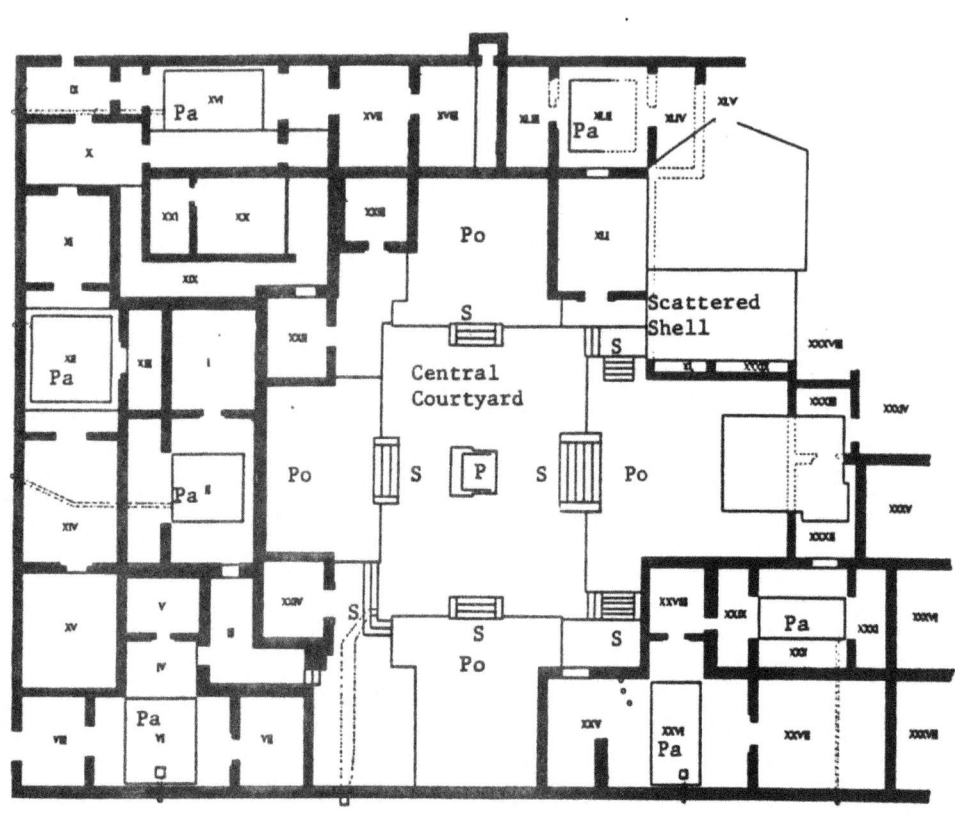

CC = Central Courtyard/Principal Patio
P = Platform Altar
Pa = Patio
Po = Portico/Porch
R = Room (designated by Roman numerals
S = Stairway

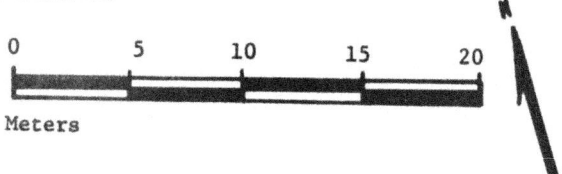

150-153; Figures 281-283, 285). Tlamimilolpa was designated Site 1:N4E4 by Millon (1973). Eight species came from the Panamanian province and eight from the Caribbean, plus one, Unio discus Lea, 1838 (a freshwater clam), from riverine habitats. The most numerous of all was the Gastropoda Conus interruptus (Conus Ximeniconus ximenes Gray, 1839 [Synonym: C. interruptus Broderip and Sowerby, 1829]) (Keen 1971:669-670), which were cut and drilled for use as pendants or "tinklers" (Linne 1942:133, 152; Figure 259). Five of thirteen burials and one of the three caches contained marine shell. I have summarized these data and adapted and corrected the Mollusca classifications to those of Warmke and Abbott (1962) and Keen (1960, 1971). (See Figure 20.)

Tlamimilolpa Burial 1

The body had been cremated in situ in a subfloor grave in Room 16 and was accompanied by over 1200 artifacts (Linne 1942:132). Included among the grave goods were eight shell ornaments, six of which were cut and pierced Oliva reticularis Lamarck, 1811, and had a maximum length of 4.4 cm. Most had been in direct contact with the crematory fire. One flat, oval pendant (1.5 x 1.8 cm), pierced at one edge for suspension, was made of Spondylus princeps Broderip, 1833.

A "great quantity" of mollusks were in association with Burial 1, including five Caribbean and six Pacific species. The Pacific specimens included: Chama coralloides Reeve, 1846 = Chama echinata Broderip, 1835; Lithophage aristata Dillwyn, 1817; Pinna maura Sowerby, 1835 = Atrina maura Linnaeus, 1758; Pteria margaritifera Linnaeus, 1758; Spondylus princeps Broderip, 1833; and Turritella cumingi Reeve, 1849 = Turritella lucostoma Valenciennes, 1832. The five Caribbean genera were: Fasciolaria (Pleuroploca) gigantea Keiner, 1840; Spondylus americanus Gmelin, 1781; Strombus pugilis Linnaeus, 1758; Thais fasciata Reeve; and Turbinella scolymus Gmelin.

Tlamimilolpa Burial 2

The interment of a "youth past childhood" in a subfloor pit in Room 22 included "numerous offerings" (Linne 1942:133). Eight shell pendants or "tinklers" made from the points of Conus interruptus Broderip and Sowerby, 1829 = Conus (Ximeniconus) nimenes Gray, 1839 were recovered. Each had been pierced for suspension (1934:133, Figure 259).

FIGURE 20: PLAN OF THE TLAMIMILOLPA SITE (1:N4E4)
(after Linne 1942:Plate 1; Millon 1973:Map 1)

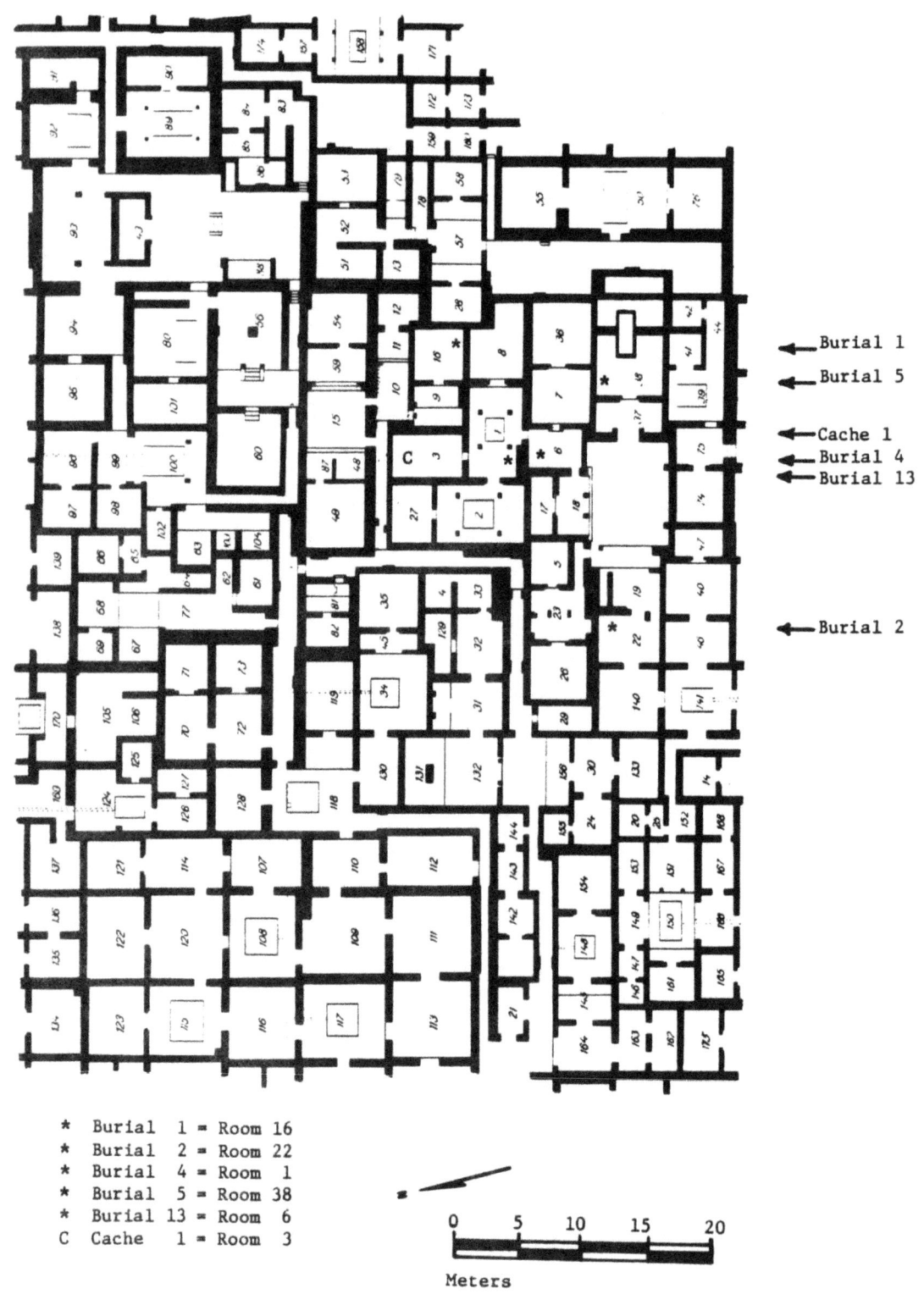

* Burial 1 = Room 16
* Burial 2 = Room 22
* Burial 4 = Room 1
* Burial 5 = Room 38
* Burial 13 = Room 6
C Cache 1 = Room 3

- 48 -

Tlamimilolpa Burial 4

The burial of an adult male in a subfloor pit in Room 1 included an unspecified number of freshwater clam shells, Unio discus Lea, 1838 (Linne 1942:135).

Tlamimilolpa Burial 5

The mortuary offerings with an adult male buried in a pit beneath the floor of Room 38 (Linne 1942:136) included two shell ornaments. One was ground into an oval shape from a piece of Spondylus americanus Gmelin, 1781, and had a small central perforation. The second was a pendant or "tinkler" shaped from Oliva reticularis Lamarck, 1811 (1942:133:Figure 259). Other mollusks present, but unspecified as to quantity, included Thais deltoidea Lamarck, 1822.

Tlamimilolpa Burial 13

An aged, unsexed adult was interred in a subfloor pit in Room 6 (Linne 1942;141). The mortuary offerings in shell included two mussel shell "rings" made from Cardium isocardia Linnaeus, 1758, which had the "convex side only partly ground smooth, closely around the excised hole runs a red painted circle" (1942:139; Figures 283, 285). There was one violet irridescent "mother of pearl" ornament shaped like a thin "spade" which was manufactured from Pinna maura Sowerby, 1835 (1942:139, Figure 282), which should be reclassified as Atrina maura Sowerby, 1835. Linne also recorded a single ceramic ornament (adorno) shaped like a small mussel, "orginally no doubt one of a series of similar ornaments attached as a border along the lower edge of some tripod vessel" (1942:139-140, Figure 281). The specimen more likely came from a censer, fragments of which were found nearby.

Tlamimilolpa Cache 1

In addition to two Cylindrical Tripod Support Vases and four Basal Break (Recurved Rim) Bowls, all dating to the Late Tlamimilolpa phase (ca 400-500 A.D.), Linne (1942:142-143) recorded "four shells of four different species" contained in one of the bowls. These included Chama coralloides Reeve, 1846 = Chama echinata Broderip, 1835; Pecten ventricosus Sowerby, 1842 = Argopecten circularis Sowerby, 1835; Fasciolaria (Pleuroploca) giganta Kiener, 1840; and Oliva reticularis Lamarck,

1811. The Fasciolaria and Oliva are Caribbean species, while the other two are Panamanian.

As at Xolalpan, the Tlamimilolpa residence had no specimens of Spondylus calcifer Carpenter, 1857. However, species from both marine faunal provinces were well represented, including Spondylus princeps from the Pacific and S. americanus from the Gulf Coast. Burial 1 and Cache 1 contained mollusks from both provinces. Subsequent published reports on marine mollusks from Teotihuacan urban residences lacked, in most instances, sufficient information to detail marine province origins (cf. Noguera 1955; Sejourne 1959, 1966a, 1966b, 1966c; Piña Chan 1963; Acosta 1964). There is, of course, the distinct possibility that the Spondylus americanus specimens were misidentified and are, in reality, S. calcifer. Malacology was less advanced as a science in the 1930s and early 1940s in comparison to today's state of the art (Keen 1971).

Tlajinga 33 Excavations

Site 33:S3W1, initially defined by personnel of the Teotihuacan Mapping Project (Millon 1973), was located in the ancient barrio of ancient urban Teotihuacan called Tlajinga, named after a nearby hacienda. This site was dated to the Early and Late Tlamimilolpa, Late Xolalpan, and early Metepec phases on the basis of ceramics, figurines and other material culture obtained during the surface survey and site mapping. William T. Sanders and his colleagues selected Tlajinga 33 for excavation because of the unusual quantities of poorly fired sherds and "wasters" which suggested that the site was a ceramic manufacturing locus specializing in the production of San Martin Orange basins or "cooking pots" (Krotser and Rattray 1980, Sanders et al 1982). The site was partially excavated by Rebecca Storey, Randolph Widmer, and Sanders in 1981, and a report on the findings at Tlajinga 33 was issued by Storey and Widmer (1982:21-97) as partial fulfillment of a National Science Foundation contract. (See Figure 21.)

In sum, they were able to determine that there were two economic specializations at Tlajinga 33 during the Early Tlamimilolpa and subsequent Late Tlamimilolpa phases. These craft activities, identified on the basis of the fine-screening of excavated soils, included the manufacture of bowls and dishes from a fine-grained travertine, and the production of "shell jewelry" (Storey and Widmer 1982:33, 39). They also mentioned in passing that a possible "shaft tomb," damaged and looted in "early times," was situated in an adjacent compound and

FIGURE 21: PLAN OF THE TLAJINGA 33 SITE (33:S3W1)
(after Storey and Widmer 1982:70-73, Figs. 1-4)

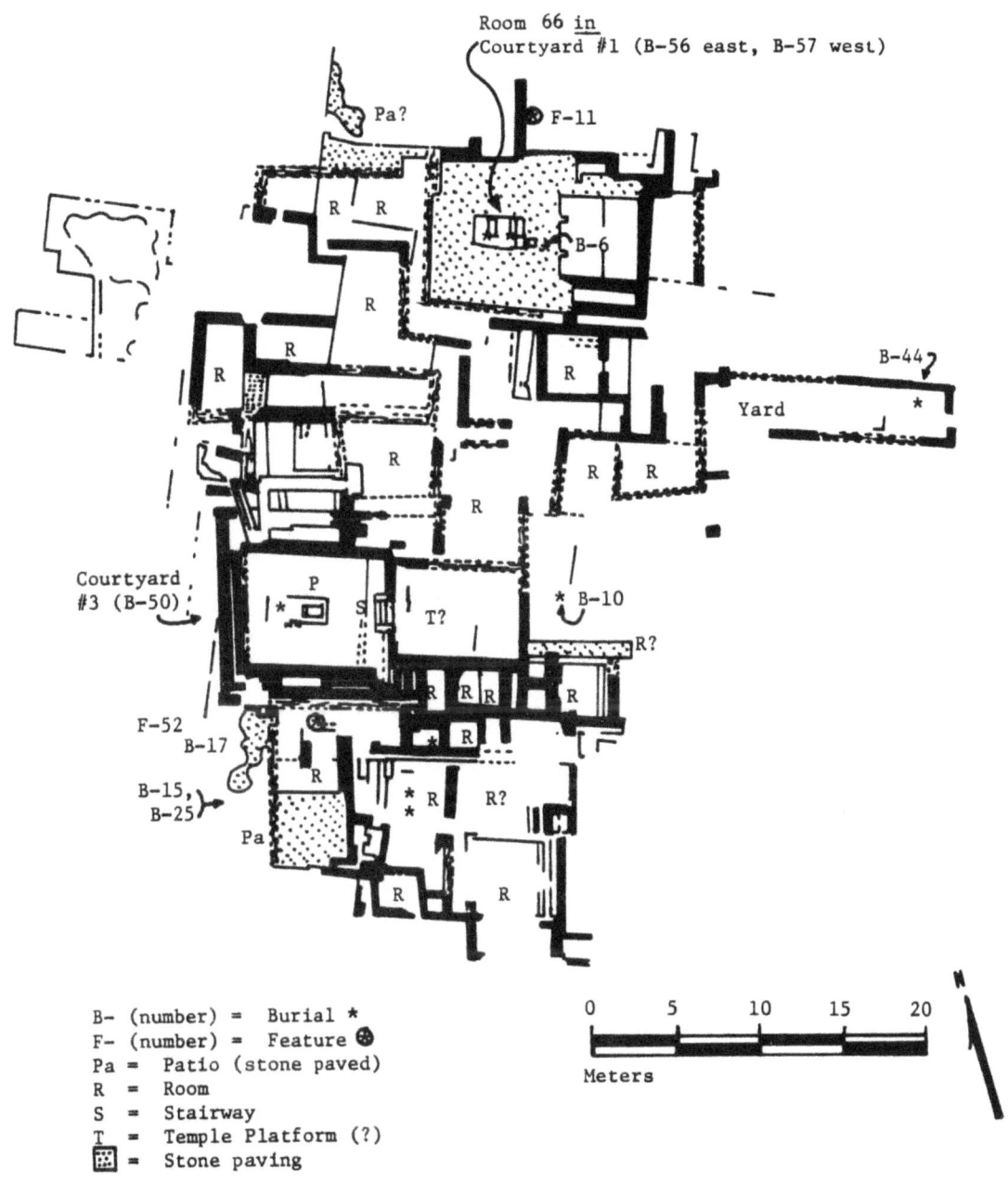

had been excavated by INAH (Instituto Nacional de Antropologia e Historia) personnel. This "tomb" contained the remnants of a ceremonial cache of ceramic vessels (unspecified wares and types), marine shell and greenstone artifacts (1982:48-49). The excavators were able to confirm that Tlajinga 33 during the Late Xolalpan and early Metepec phases was an area of San Martin Orange ceramic specialization. Open "kilns," malformed ceramics, "wasters," and kiln furniture (large sherds used to cover the open-air firing) were noted.

Tlajinga 33 covered an area of 2,280 square meters, of which 1,250 square meters or about 55% was excavated through several construction phases to a sterile base. This apartment compound was "loosely organized" in spatial form and reminded Storey and Widmer of the sprawling, irregular multi-roomed Tlamimilolpa site excavated by Sigvald Linne (1982:25-26, 52, 66-68). A total of seventy burials (109 individuals comprising fifty-five adults, eight adolescents, eleven juveniles, thirty-four perinatals, and one unidentified interment) were excavated, as were sixty-seven features (1982:33, 54, 60, 80-97). Ten burials (fourteen individuals -- predominantly adults) and only two of the features contained shell artifacts or marine shell fragments. Storey and Widmer reported that Early Tlamimilolpa primary burials had a "jade" or marine shell bead placed in each individual's mouth, a mortuary practice which was discontinued by Late Xolalpan times. In addition to marine shell, primary burials also contained ceramic vessels, obsidian blades and other tools, and greenstone offerings (ofrendas) as grave furniture. Those burials and features with shell offerings included:

Tlajinga 33 Burial 6

This adult, primary interment dated to the Early Tlamimilolpa phase, and was located beneath the floor of Courtyard 1. Grave goods included two small ceramic "jars" and one plain and one notched shell bead (Genus and species not identified).

Tlajinga 33 Burial 10

A perinatal, primary burial located beneath the floor of Room 74 which dated to the Early Tlamimilolpa phase. Mortuary offerings included four obsidian blades, two "outcurved" rim bowls (not further identified), and one copy of a Thin Orange bowl. The only marine shell offering consisted of "one shell outcurving bowl" (1982:81), not further detailed as to possible Genus or species.

Tlajinga 33 Burial 14

This adult, primary interment, located beneath the floor of Room 92, dated to Late Tlamimilolpa times. The mortuary furniture included sixteen pottery vessels (not further delineated) and two shell beads (1982:81). The Genus and species were not discerned.

Tlajinga 33 Burial 15

This was another adult, primary burial also dating to the Late Tlamimilolpa phase, and was recovered from a pit beneath the floor of Room 33. Grave goods included one bone needle, five ceramic vessels (one jar, one bowl and three matte dishes), and one unidentified "small shell."

Tlajinga 33 Burial 17a

The burial, located beneath the floor of Room 32, dated to the Late Xolalpan phase. Notably, this was the only child interment and only secondary burial to have a marine shell offering. The singular mortuary object was one unidentified shell fragment.

Tlajinga 33 Burial 25

Also located beneath the floor of Room 33 was an adult, primary burial again dated to the Late Tlamimilolpa phase. The mortuary offerings included sixteen ceramic vessels, one of which was a _copy_ of a Thin Orange bowl, a ground stone "toy" mano and _metate_, and one "shell disk" (1982:83). The marine artifact was not further identified.

Tlajinga 33 Burial 44

This adult, primary interment dated to Late Tlamimilolpa times was found beneath the floor of Room 53. The only grave good was one "shell bead," Genus and species not delineated.

Tlajinga 33 Burials 50a through 50e

The interment consisted of five adult, primary burials beneath the floor of Courtyard 3. Mortuary furniture included ten ceramic vessels, thirty obsidian blades, one mica disk, and one "shell pendant" (1982:86). The morphological characteristics of the pendant, and its Genus and species were not further explicated.

Tlajinga 33 Burial 56

Dating to the Early Tlamimilolpa phase, this adult, primary burial was recovered from beneath the floor of

Room 66. The grave goods consisted of thirty-four ceramic vessels (including one Florero and one Cantaro). In the main, the pottery vessels were "outcurving" rim bowls, one of which contained three "seashells" as part of the offering (1982:86). No further data was provided regarding the ceramics or marine shell.

Tlajinga 33 Burial 57

This interment consisted of an adult, primary burial located below the floor of Room 66. The grave goods included three ceramic vessels and a variety of marine shell artifacts: three small shell pendants, two "shell disks carved in rattlesnake form," two "shell filagree disks," and three fragments of shell. Burial 57, a male, was "without a doubt, the highest ranked individual in the compound at this time [Early Tlamimilolpa] and probably for all phases of the compound. ... he was wrapped in a cloak or robe which was made or embroidered with approximately 4,000 olivella shells, some of which seem to have traces of paint on them. More impressive is the shell headdress which he was wearing. This consisted of a pair of large ring-shaped carved rattlesnakes on both sides of the head. Below these were suspended thin shell nacerous disks decorated with an excised delicate slot pattern" (1982:34). The significance of this burial also required that a "shrine room" was razed and a carved stone Huehueteotl be ritually broken. Olivella are a widely distributed shell found in both marine faunal provinces. It was likely that the shells comprising the cloak or robe were Olivella mutica Say (Genus Olivella Swainson, 1846) from the Caribbean Marine Faunal Province, rather than the larger Oliva (Oliva) reticularis Lamarck, 1811, the "Netted Olive," which is a close biological relative (Morris 1951:173, 214).

Tlajinga 33 Feature 11

This feature consisted of a broken Early Tlamimilolpa censer base which had been inverted over a layer of incompletely burned charcoal and two marine shells. One was a Busycon perversum Linne (Genus Busycon Bolton, 1798), while the other was a Spondylus calcifer Carpenter, 1857. The Busycon contained the fragments of a "retracted hermit crab" which later "fell out after excavation" (Storey and Widmer 1982:34). This species of Busycon, the "Lightning Conch," is a Gastropoda of the Caribbean Marine Faunal Province (Morris 1951:76, 205-206). It is a brightly colored, relatively rare shell, and in older specimens the colors fade and the shell becomes a dull white.

Tlajinga 33 Feature 52

This refuse pit in Room 92 was located near Burial 32. The pit contained obsidian blades and a large amount of charcoal, plus one "shell fragment" (Genus and species unidentifiable.)

Noguera and Sejourne's Excavations

In 1955 a subfloor interment with substantial grave goods was excavated in a mound (tlatel) located north of the Rio San Juan and south of the old highway to San Martin de los Piramides (Noguera 1955). This site was in Square N1E2 (Millon 1973) but its precise location is now uncertain so that it is referenced as the "Rio San Juan Tlatel at Highway." Associated with the burial were twenty-two ceramic vessels, six Pecten -- one peforated -- (described as "Chamidos") and a pendant derived from the columnella of a Strombus (Noguera 1955:43, 46, Figure 2). The pottery dated to the Late Tlamimilolpa and Early Xolalpan phases, ca 400-600 A.D. (Kolb 1965a).

Laurette Sejourne's excavations at the Zacuala Palace or apartment complex, Site 3:N2W2, during the years 1955-1958, produced one hundred marine mollusks (1959:124). Burial 1 had two large ovoid shell ornaments (1959:69, Lam. 38). Burial 27 contained ten cut Spondylus shell discs, ten shell beads, and nine rectilinear pieces of cut shell, while "el craneo tenia un burrete hecho de placas de concha roja fuertemente quemadas" (1959:64-68, Figure 45B, Lam. 37). The mollusks were not specifically identified but were said to have come from both the Pacific and Gulf coasts. Of six shells illustrated, two were Oliva Bruguiere, 1789 and one was Fasciolaria Lamark, 1799 (Warmke and Abbot 1962:119, 121-122; Keen 1971:610-611, 622-625).

Sejourne's 1963-1964 excavation report on the Tetitla urban residence (Site 1:N2W2) stated that some of the interments and caches contained marine shells. Burial 11 contained two Spondylus shells, and an ofrenda had two Spondylus and one unidentified Gastropoda (Sejourne 1966a:Lams. 54, 58). Sejourne (1966a:233, 236-237, Figures 214, 216 Row 3:1-4, Lam. 64) noted other mollusks, including twenty-three different shells ("conchas y caracoles") from burials, and a total of 480 shells from the Central Courtyard debris. None of these latter mollusks were further identified except that she determined that the shells came from both the Pacific and "Atlantic" coasts (1966a:236, Lams. 54, 58, 64; 1966c:160-165, Figures 111A-113, Lams. 43-48). Undoubtedly she meant the Caribbean Marine Faunal Province rather than the "Atlantic." At least eighteen Spondylus can be identified from the illustrations, but

Pecten and Oliva were also represented in Tetitla ofrendas. One Strombus or Turbinella specimen had been cut and drilled (1966c:164, Lam. 48).

A collar or necklace of some twenty strands of Dentalium semipolitum shell beads (each bead 2.0-3.0 cm long) was recovered at Tetitla, apparently from an ofrenda (1966c:165, Lam. 49). Dentalium (Graptacme) semipolitum Broderip and Sowerby, 1829 ranges from southern California through the Gulf of California to Costa Rica at depths of 2.0-4.5 m (Keen 1971:886). These "tusk" shells are naturally up to 30.0 mm long and have a 3.0 mm diameter at the aperture. There are no other reports of Dentalium being found in any other Basin of Mexico Classic or Postclassic sites. Only at the Barton Ramie, Belize site were Dentalium texasianum cestum Henderson reported in an archaeological context (Andrews 1969:38; Richards and Boekelman 1937:166-167). Unfortunately, Sejourne did not specify the locations of the burials and ofrendas at the Zacuala Palacio or Tetitla residence.

Other Mollusks at Teotihuacan

Sections of the Great Compound were initially surveyed by Bennyhoff and Kolb in 1964 when they were associated with the Teotihuacan Mapping Project (Kolb 1964). At that time we observed numerous fragmentary specimens of Spondylus calcifer Carpenter, 1857 on the surface within the complex, especially in the central and east central segments, since designated as Site 1E:N1W1 (Millon 1973:18-20). Additional investigations in the maguey fields in Square N3W2 immediately north of the Yayahuala and Zacuala Palacio sites also revealed quantities of Spondylus calcifer fragments and those of Gastropoda, possibly Strombus (Kolb 1973a). No complete shells were found and one human burial excavated in 1962 at La Ventilla A (Site 1:S1W2) contained over one hundred marine shells, both Gastropoda and Pelecypoda (Piña Chan 1963:52). Unfortunately, to my knowledge, no further classification has been published. A most unusual ceramic quetzal bird effigy vessel was excavated by Juan Vidarte de Linares at La Ventilla B (Sejourne 1966a:Lam. 37-39). The Site (1:S1W3) was destroyed by the landowner in 1964. This globular vessel with a simple tecomate-like rim apparently had a series of actual Oliva shells as decorations attached along the rim. There appeared to be four Spondylus fragments represented in one human interment (enterro del posillo) from the Palace of Quetzalpapalotl, Site 2:N4W1 (Acosta 1964:36, Figure 58). (See Figure 15.) Spence (1971, 1976) has recorded no other urban burials as having marine mollusks as grave goods.

CHAPTER FIVE

SANTA MARIA MAQUIXCO EL BAJO: TC-8

Introduction

In 1960, a large Classic Teotihuacan Period village site was located during a preliminary archaeological site reconnaissance of the Teotihuacan Valley, a northeastern segment of the Basin of Mexico, during the initial phase of the Teotihuacan Valley Project (The Pennsylvania State University, William T. Sanders, director) (Sanders 1965:109). Called Santa Maria Maquixco el Bajo, to differentiate this site from the Santa Maria Maquisco el Alto locality, the multicomponent Maquixco site was ultimately designated TC-8 (Teotihuacan[Valley] Classic [Period Site] number eight). The Maquisco el Alto site was situated in the Cerro Gordo North Slope Ecological Zone, whereas the TC-8 Maquixco el Bajo site was located in the Lower Teotihuacan Valley Ecological Zone. During the 1961 and 1962 field seasons at TC-8, five excavations were conducted, in the main, to obtain artifacts so as to be able to develop a ceramic and figurine relative chronology, and provide data on rural Classic Teotihuacan intrasite settlement patterns and architectural styles (Kolb 1962, 1963, 1965a, 1970, 1972b, 1973b, 1979:378-386, 556-563; Marino 1965:108, 147, 164, 169, 1975:303; Sanders 1965:109-114, 1966; Santley 1984:52, 54, 60-61). (See Figure 22 and Appendix V.)

The excavations in 1961 included the trenching and sectioning of a low temple pyramid (TC-8:Pyramid) and the excavation of a portion of an apartment complex (TC-8:1). The 1962 field season involved the continued excavation of the complex and a contiguous apartment complex (TC-8:1-2), the complete excavation of an apartment complex (TC-8:4), and the partial excavation of a large complex and patio in an adjacent mound (TC-8:3). These excavations were undertaken by Sanders, Maurice Mook, Barbara Price, Thomas Krajci, and Charles Kolb. (See Figure 23.)

During the 1963 field season most Project activities were devoted to settlement pattern surveys, and included the mapping of Classic Teotihuacan sites especially in the Delta, Lower, and Middle Valley Ecological Zones of the Teotihuacan Valley. A field team led by Kolb, with the assistance of Ira Smith, III and R. Brooke Thomas, prepared a grid, mapped the entire TC-8 site, and collected artifact samples from each of seventy-three

FIGURE 22: PLAN OF THE ARCHAEOLOGICAL REMAINS AT THE SANTA MARIA MAQUIXCO EL BAJO SITE (TC-8)

(after Kolb 1963:Map TC-8; Sanders et al 1979:335, Fig. 8.21)

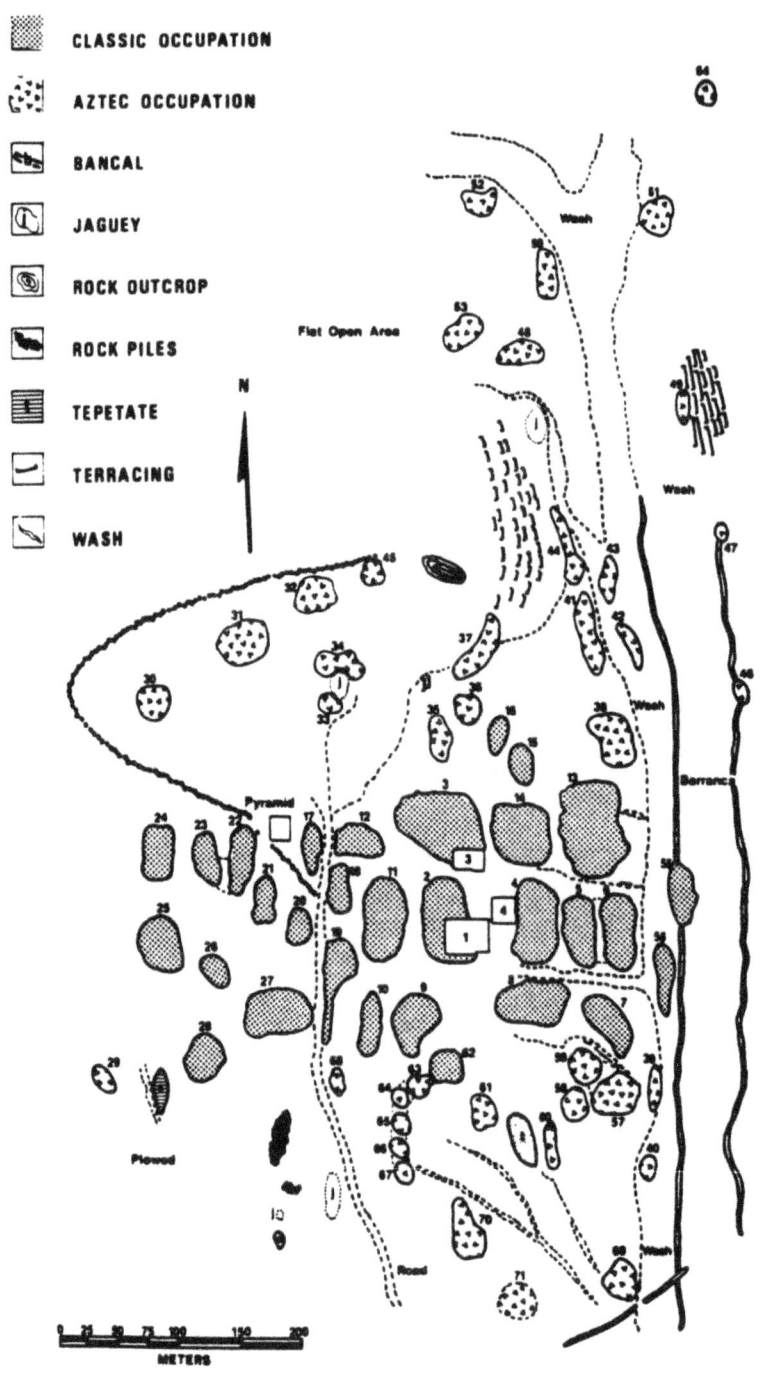

Mounds are identified by number

- 58 -

FIGURE 23: PLAN OF THE SANTA MARIA MAQUIXCO EL BAJO SITE
EXCAVATIONS (TC-8:1-2, 3, 4)

(after Kolb 1963:TC-8:1-4 Plan; Sanders et al 1979:338, Fig. 8.22)

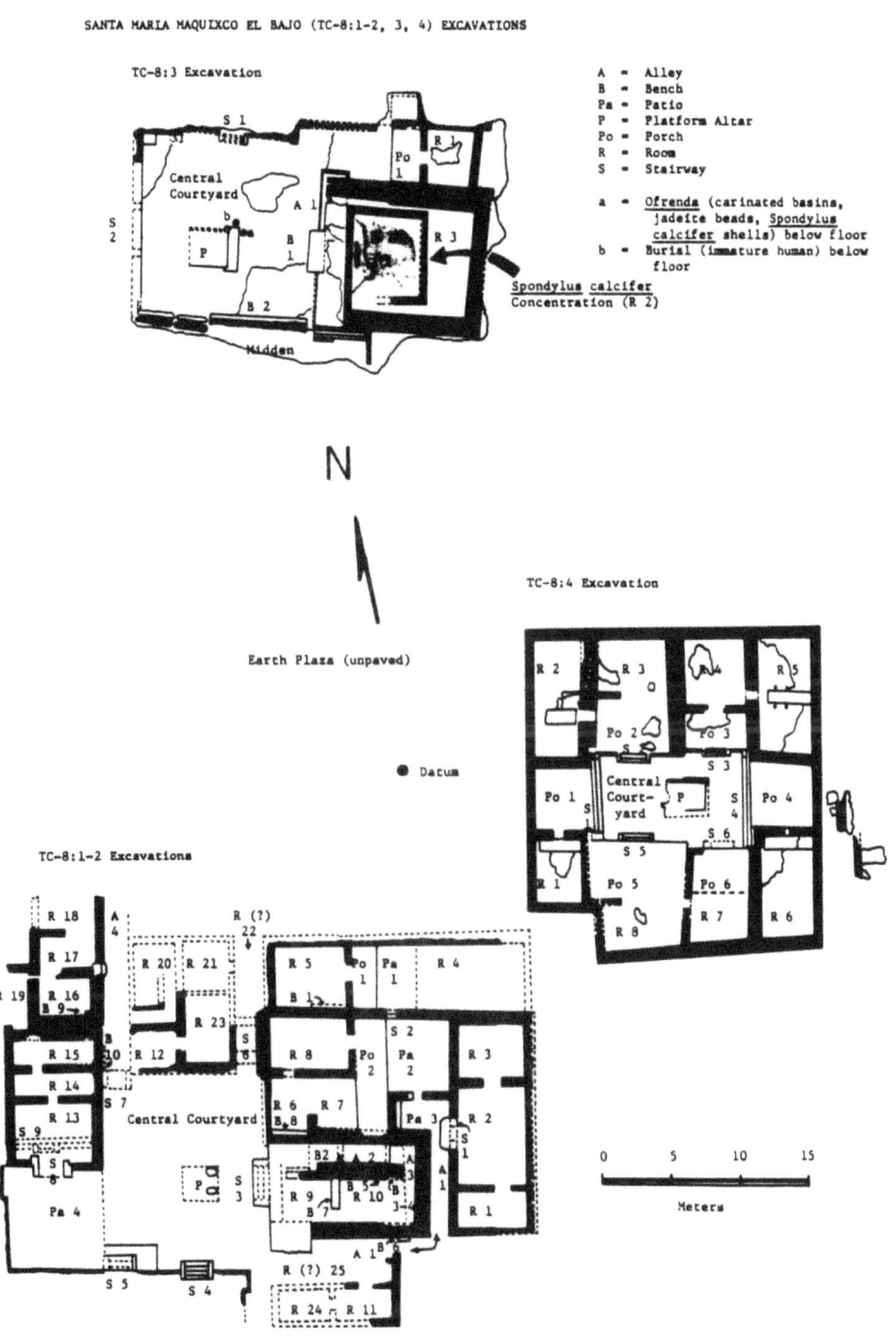

mounds (tlateles) (Kolb 1963, 1965a, 1965b, 1970, 1973b, 1979a:378-386, 556-563). The laboratory analysis of Classic site artifacts began in 1962 and continued through 1968.

At the 1965 Teotihuacan Mesa Redonda, Sanders (1966:123-148) presented a paper, "Life in a Classic Village," in which he attempted to reconstruct the nonmaterial aspects of culture and life in the TC-8 rural community during the Xolalpan phase. He noted that the laboratory phase of research was "still in progress" so that his conclusions were "tentative and incomplete" (1966:123). Subsequently, Robert Santley used TC-8 data for a comparison with his Loma Torremote Late Formative village (1977:89-216) and Kolb prepared a final settlement pattern analysis of the TC-8 site (1979a:378-386, 556-563). Sanders, Parsons, and Santley (1979:334-354) published a revised version of the 1966 report as "Maquixco Bajo: A Middle Horizon Village." The natural ecology and contemporary agriculture were summarized by Sanders et al (1970). The following descriptive summary of the Santa Maria Maquixco el Bajo site draws upon all these aforementioned works.

The Prehistory of Maquixco el Bajo

The TC-8 site, also called Santa Maria Maquixco el Bajo, or Los Tres Reyes Yzquitlan (Izquitlan), or Loma de Calaveras, is located at the base and on the lower flank of a small hill locally called Loma or Cerro de Calaveras. Topographically, the site is located on the gently sloping northern piedmont of the Lower Teotihuacan Valley's Lower Piedmont Ecological Zone and lies 1.5 km north of the Lower Valley Alluvial Plain. The site is situated between the 2,305-2,320 m contours above the Plain which begins at the 2,270 m contour. Soils in the site area have a sandy to loamy texture and are tan to light brown in color (lama amarilla) with intermittent patches of darker humus (tierra negra). There is moderate erosion in the eastern section of the site, where soil depths range from absent to at least 130 cm, the former where tepetate is exposed because of sheet and small gulley erosion. The southern portion of the site is seriously eroded and a number of washes have formed. A canalized barranca of uncertain date is found at the eastern edge of the site, and areas of serious sheet erosion are located north of the site area. Current (1960-1962, 1970, and 1972) vegetation in the vicinity includes nopal (Opuntia ssp.), huizache (Acacia farnesiana), maguey (Agave ssp. usually A. americana), and the non-indigenous pirul (Schinus molle, the Peruvan "pepper tree"), in addition to various grasses.

No modern cultural features were originally associated with the site, but in 1970 and 1972 I noted that the east central segment of the site was being utilized as garbage dumps for the Municipio de San Juan Teotihuacan and the modern villages of San Juan Evangelista and Santa Maria Maquixco. There was also evidence of clandestine excavations ("pot hunting") in six of the larger mounds (tlateles) in 1970 and 1972. The site area in 1960-1964, 1970, and 1972 was used in the main, for the grazing of sheep, goats, and some cattle, but a maguey plantation with terraces or stone bancals (earth ridges reinforced with stone, built parallel to the slope angle) is found in the northern and central site areas. These bancals and three small jagueys (rock-cut water reservoirs) within the site are of unknown date but were most likely constructed after the Classic Period, probably during the early Colonial era when Hacienda de la Cadena was established.

The total multicomponent site occupies 36.0 ha, while the Classic Period component occupies minimally 8.0 ha and maximally 10.5 ha. Seventy-three mounds were identified of which fifty-three and the Pyramid had at least some Classic occupation in the form of architectural remains, ceramic figurines and sherds, and lithic artifacts dated to the Classic. The site had five phases of the Preclassic, all six phases of the Classic, and four phases of the Postclassic represented along with traces of early Colonial occupation. Six associated non-Classic sites included TF-138 (Teotihuacan Formative), TT-133 (Teotihuacan Toltec), and TA-28, TA-219, TA-220, and TA-221 (Teotihuacan Aztec). During Postclassic Aztec times part of the Classic site area was overlain by an Aztec line village (TA-219/221). To the west is TC-7, while Classic sites TC-11, TC-12, and TC-121 are located immediately to the east. In spite of the Preclassic and Postclassic occupations, the Classic site was easily identified because of the architectural and artifactual remains, the size of the mounds, and evidence of a formal plan with building orientations similar to those of the Supra-Regional Center.

The Classic site component had a moderate to heavy distribution of lithic materials including 8,121 obsidian blades, 213 cores, 92 projectile points, 869 irregular flakes, and 255 bifacially worked scrapers (Sanders et al 1979:348-349). Some chunks and cobbles of obsidian were also noted. Ground stone tools, especially mano and metate fragments and mortar and pestle fragments as well as more specialized lithic materials, including artifact fragments of serpentine, flint, and quartzite were also found. Evidence of manufacturing debris was not discerned during excavation or in soil samples subsequently analyzed. The excavations and surface surveys produced over 145,000 sherds of which 15,432 diagnostic Classic Period sherds (9,784 rims and 5,648 bodies) were retained for intensive analysis Kolb (1965a, 1965b). A total of 2,290 ceramic figurine fragments (2,150 from excavation and 140 from survey were also recovered and studied (Kolb 1970, 1973b; Hodik 1973).

Originally, the TC-8 site was interpreted by Marino (1965:108, 147, 164, 169) as a compound village-town with quadrangular arrangements of multiple family houses dating to "Middle and Late Teotihuacan," i.e., Late Tlamimilolpa through Metepec phases, based on the preliminary analyses. The site was also stated to be one of his "aligned east-west" sites of "Middle to Late Teotihuacan" associated with a cruciform grid and sites TC-7, TC-25, TC-87, and TC-119. In discussing the Toltec (TT-133) component of TC-8, Marino (1975:303) considered it to be a dispersed low-density Mazapan phase village. Sanders (1965:104, 107-116, 120-121; 1966:123-125, 133, 140-142; Sanders et al 1979:336, 353-355) considered TC-8 to be an excellent example of a Classic village dated to the Xolalpan phase and characterized it a corporate community containing the houses of lineages which had nuclear family apartments. The site was the socioeconomic equivalent of the Aztec/Culhua Mexica calpulli (Kolb 1979a:341-345, 353), and classified as a Large Nucleated Village with 140-150 families or 500-600 people during its apogee in Xolalpan times, or in the newer terminology Middle Horizon (Sanders 1966:126, Sanders et al 1979:336). Flannery (1974b:164) has observed that Nucleated Villages lacked "large-scale ceremonial-civic architecture," as did TC-8.

I have proposed that the site grew from a dispersed Hamlet of twelve small dwellings during the Late Preclassic (ca 50 B.C.-150 A.D.), to a Small Nucleated Village during the Early Classic (ca 150-450 A.D.), to a Large Nucleated Village during the Middle-Late Classic (ca 450-650 A.D.), and became a Small Nucleated Village during the terminal Late Classic Metepec phase (ca 650-750 A.D.) (Kolb 1979a:386). I suggested a maximum population of 773 during the Xolalpan phase (ca 500-650 A.D.), a figure somewhat greater than Sanders' 600 estimate (Kolb 1979a:560). Subsequently, Early Postclassic Toltec occupations included a Xometla phase Hamlet and quite small Mazapan phase Small Nucleated Village (ca 800-900 and 900-1150 A.D.), eg. Phase One and Two of the Second Intermediate. The final significant occupation was during the Late Postclassic Aztec/Culhua Mexica (Aztec II-IV) phases, eg. Phase Three of the Second Intermediate and Late Horizon, when an Aztec line-village occupied the northeastern section of the TC-8 site from ca 1200-1519 A.D. However, Aztec Black/Orange wares on which the chronology was based, are now known to persist to at least 1650 and as late as 1720 in regions of the Basin of Mexico (Parsons 1966, Charlton 1972).

Based on the continuation of the Aztec IV Black/Orange ware, Majolica ceramics, and a bronze oval religious medallion of the Holy Family (Joseph on the obverse and Mary and the Infant Jesus on the reverse) it was clear that the Maquixco site was occupied into the early Colonial period. The medallion (Specimen 8249) was of a type produced during the sixteenth century in Spain for use by priests of the Dominican and Franciscan orders. Some sherds of Aztec Black/Orange bore fragmentary designs apparently representing the Austrian "double eagle," a motif also used on Iberian ceramics of the sixteenth century.

The Colonial and Recent History of Maquixco el Bajo

Archival and cartographic sources were examined in order to discern the existence of the Santa Maria Maquixco el Bajo site during the early Colonial eras. There were numerous documents concerning Maquisco el Alto, the community located in the Cerro Gordo North Slope Ecological Zone which was an estancia in the cabecera of Tepexpan in the corregimiento of Teotihuacan in 1530 and 1580 (Archivo General de Indias, Justicia, 1530 (208 4 :3-7), Paso y Troncoso 1580 (1905 [6]:209), Gibson 1964:48-49, 86, 110). Francisco de Casteñada's official reports sent to King Philip II of Spain and the Council of the Indies, dated February 22-23, 1580, included information about Teotihuacan Valley communities (Paso y Troncoso 1580(6):Mapas, Gamio 1922b:Lams. 138-139, Nuttall 1926:Pl. 1). Neither the "Mapa de Tecciztlan, Acolman, Teotihuacan, y Tepechpan, Corregimiento de Tecciztlan, en 1580" or the map of "El pueblo de San Juan Teotihuacan y sus monumentos en 1580" illustrated Los Tres Reyes Izcuitlan or Santa Maria Maquixco (el Bajo) or Hacienda de la Cadena. Nor was there any depiction of any community between Tepechpan (Tepexpan) and Teotihuacan in the region where Maquixco existed. Interestingly, both maps illustrated the ceremonial complex of the ancient urban center of Classic Teotihuacan as "Oraculo de Montecuma." Casteñada's report did, however, indicate Tres Reyes Yzquitlan as one of twenty-seven dependencies of Acolman (Nuttall 1926:60). Maquisco el Alto was referenced as Santa Maria Maquiteco one of twelve dependencies of Tepechpan

(1926:59) and later as Maquixco, one of the dependencies of Tepechpan (1926:78).

Therefore, the archaeological Santa Maria Maquixco el Bajo (TC-8), Tres Reyes Yzcuitlan (Nuttall 1926:60) or Tres Reyes Izquitlan (Gamio 1922b:382-383, 389) which existed in 1580 ceased to exist by 1595. A new pueblo called Maquixco in 1595 was located at the edge of the alluvial plain immediately east of the Hacienda de la Cadena, and both were within the jurisdiction of Texcoco (Archivo General de la Nacion, Tierras, 1595 (1520[2]:6, Mapa; Gamio 1922b:389, Lam. 141; Gibson 1964:88-89). However, a "Plano del pueblo de San Juan Teotihuacan en 1585" indicated "Los Reies sujecto a Aculma" at a location west of the Barranca de Malinalco, consistent with the TC-8 site location.

The evidence suggests that TC-8 as Los Tres Reyes/Los Reies existed until about 1585-1595 after which the modern Santa Maria Maquixco el Bajo was established 1.5 km southeast of the slopes of Cerro de Calaveras. I believe this resettlement between 1585-1609 was the result of the establishment of Hacienda de la Cadena at the edge of the alluvial plain. In 1609 the hacienda was called venta Cadena and had Capitan Don Diego Velasquez de la Cadena, Caballero de la Orden de Santiago, as proprietario (Gamio 1922b:382-383, citing an untitled Ms. in the Archivo Municipal de San Martin de los Piramides). With los Indios living in close proximity to the hacienda, or within its walls, the archaeological sites (TF-138/TC-8/TT-133/TA-28 and TA-219/TA-220/TA221) were uninhibited and the area converted into maguey cultivation and probably pasture for the hacienda livestock. A thorough search of the documents and maps in the Archivo General de la Nacion by Thomas Charlton and Charles Kolb in 1963-1964 yielded no further evidence on the archaeological Los Tres Reyes. Subsequent studies of other documents (Fernandez Navarrete 1842-1895, Garcia Icazabalecta 1886-1892, Paso y Troncoso 1905-1948) produced no relevant post-1595 data.

Hacienda de la Cadena and Maquixco continued in existence as dependencies of Acolman (Aculma) into the 1800s (Castillo 1922:706, 709, 710). Santa Maria Maquixco was a barrio of the Municipio of San Juan Teotihuacan before 1852, when it became a pueblo. In 1849 it was spelled Maguisco, but delineated as Maquixco in the 1852 and 1877 censuses (1922:710). "Hacienda Cadena" was a dependency of Acolman from 1870-1900, and since 1900 a dependency of San Juan Teotihuacan. Gamio (1922c:477) published census data for both Maquixco and Cadena for the years 1868, 1869, 1877, 1879, 1894, and 1901. In the latter year the pueblo had 305 inhabitants while the Hacienda had 191. The Hacienda ceased to function after the Mexican Revolution, 1910-1920, and has been in disrepair since then. Some inhabitants from the <u>pueblo</u> of Santa Maria Maquixco have lived in the old <u>hacienda</u> buildings since that time. Marquina (1922:589-591) provided a detailed architectural plan of Hacienda de Cadena as of 1920.

Xolalpan Phase TC-8

The TC-8 archaeological site reached its greatest spatial extent and highest demographic level during the Xolalpan phase (ca 500-650 A.D.) of the Classic Period or Middle Horizon. The Classic village occupied an area approximately 250 m north-south and 500 m east-west, the latter dimension at a right angle to the slope of Cerro Calaveras. The western (upper) one-third of this rectangular area lay on a flat plateau-like surface and contained ten of the smaller Classic mounds and the Pyramid adjacent to a 30.0 by 40.0 m plaza and other open areas. It was inferred that these structures were civic or public buildings (Sanders 1966:125, 126; Sanders <u>et al</u> 1979:334, 336) occupying an area of about 3.0 ha. (See Figures 22 and 23.)

The Temple Pyramid

The "village temple" pyramid was trenched in 1961 but was "rather unimpressive" but did argue for a well-integrated corporate community (Sanders 1966:139, Sanders <u>et al</u> 1979:352). The badly pitted temple platform had been pitted and looted perhaps in Prehispanic times and was damaged by locals at night during the 1961 field season. It was apparently built in two stages, with the more recent stage an expansion and renovation. The final construction consisted of a single <u>talud</u> and <u>tablero</u> facaded with a balustraded staircase facing south toward the plaza. the original height was 1.3 m and basal dimensions of the final construction stage were 10.4 by 14.6 m (151.84 m^2). No <u>ofrendas</u> or ceremonial middens were found, and the quantities of ceramic figurines and ceramics were rather small. Fragments of Classic composite censer vessels (Late Tlamimilolpa through Late Xolalpan phases were discerned, and only one sherd from a copoid vessel, a potential ritual or ceremonial pottery, was recorded (Kolb 1987a, 1987b). No marine or freshwater shells were recovered during the excavation or subsequent survey (Kolb 1973a).

Structure TC-8: 1-2

Excavations were also conducted in mounds TC-8: 1-2 during 1961 and 1962 (Sanders 1965: 111-112, 1966: 127-128; Sanders et al 1979: 337; Kolb 1979a: 378-379). Initial survey in 1960 suggested these were separate structures, but the excavations revealed them to be segments of a single apartment complex with an open Central Courtyard or large paved patio between them. The maximum extent of debris was 2,730 m^2, but the actual structure covered approximately 1,500 m^2, of which 900 m^2 was excavated. The structure consisted of up to fifteen one- to three-room apartments, often with associated porches and patios, surrounding the central courtyard. Each apartment presumably housed a nuclear family, hence from sixty to eighty persons per complex (Kolb 1979a: 379, 560; Sanders et al 1979: 337).

The arrangement of apartments, courtyard, small courtyard platform altar, and larger platforms was suggestive of a smaller version of the urban Teotihuacan residential complex of Yayahuala (Sejourne 1959). However, TC-8: 1-2 was less than half the size of the urban complex, had smaller rooms and patios, had a much more irregular plan, and was constructed of inferior materials in comparison to Yayahuala. One apartment porch had small fragments of a simple geometric design wall mural painted in red, greenish-blue, and black. The Tlamimilolpa urban Teotihuacan residence also had similar geometric wall murals in Rooms 7 and 18, with traces of murals in six other rooms of 176 excavated (Linne 1942: 115-116, Figure 190). The platform altar in the central courtyard at the Xolalpan residence had traces of red painting but no murals (Linne 1934: 48). The courtyard contained a significant concentration of irregular obsidian cores, possible evidence of the manufacture of flakes for processing maguey fiber (Sanders et al 1979: 349).

The fragmentary remains of seventeen humans were found in TC-8: 1-2, of which ten (six adult and four immature) were associated with Aztec components, three (one adult and two immature) were probably Aztec, and four (one adult and three immature) were definitely Classic Teotihuacan based on associated Late Xolalpan ceramic grave goods (Bilharz 1972, Kolb and Bilharz 1972). Aztec remains were normally buried in alleys or outside of the residential complex walls, while Classic burials were found in subfloor pits of interior apartment rooms.

The artifactual debris in TC-8: 1-2 included quantities of mammal bone: deer (Odocoileus virginianus), the American antelope (Antelocapra americana), domestic dog (Canis familiaris), cottontail rabbit (Sylvilagus, sp.), and hare (Lepus sp.). Also represented were turkey (Meleagris gallopavo), duck (Anatidae sp.), and various birds (Aves indet.), with rare freshwater fish (Pisces sp.). Most of the bone showed evidence of butchering and/or cooking associated with food processing. Animal bones tended to be concentrated along the east side of the open courtyard, in the alleys between apartments, and in rooms considered to be kitchens on the basis of associated culinary ceramics. Scattered in no apparent concentrations were whole and fragmented marine mollusk shells (n = 215), predominantly of Spondylus calcifer and one tubular bead of the same raw material (Kolb 1973a). Eight Copoid Ware vessels (three Copas cups, three Cylindrical Vases, one Tapered Vase, and one Deep Cup) were found in rooms inferred to be kitchens or on stairways leading to the central courtyard (Kolb 1987a, 1987b). (See Figures 24-28 and Appendices I-V.)

Structure TC-8:3

Mound TC-8:3 was excavated during the 1962 field season (Kolb 1962, 1979a:378-379; Sanders 1965:112-113, 1966:128, Sanders et al 1979:337). The maximum extent of debris was 3,650 m^2 but only 450 m^2 of this extensive mound was excavated. In the main, the excavation uncovered a courtyard at the south central edge of the mound, and an apartment and what I interpret as a "storeroom" (bodega) in the southeast corner of the structure. The courtyard was similar in size and construction to that of TC-8:1-2. To the north and west, staircases led to unexcavated apartment complexes above the lower courtyard. I estimate that the TC-8:3 structure may have been as large as 2,000 m^2 and only slightly smaller than unexcavated TC-8:13 (maximum debris extending 3,285 m^2) and TC-8:4 (maximum debris extending 2,595 m^2). The TC-8:3 courtyard contained a small central platform altar which had portions of a talud and tablero construction remaining.

The architectural style and construction of the excavated sections of TC-8:3 were similar to TC-8:1-2, but the "offset" rather than central courtyard suggested an overall plan dissimilar to most urban apartment compounds (Atetelco, Tetitla, Tlamimilolpa, Xolalpan, Yayahuala, Zacuala Palacio and Patios), but reminiscent of La Ventilla A System II and La Ventilla B in the urban zone (Piña Chan 1963, Kolb 1964, Millon 1976:Figure 13).

No wall mural remnants were found in TC-8:3 although some walls bore traces of specular red hematite paint. Neither were there any unusual frequencies of basalt, ceramic or other artifacts, except for concentrations of irregular obsidian flakes and bifacially flaked obsidian scrapers in the central courtyard. The fragmentary remains of thirteen humans were represented in the osteology, including two definite Aztec individuals (both immature), six probable Aztec (four adult and two immature), and five definite Teotihuacan burials (three adult and two immature) with Early to Late Xolalpan ceramics associated (Bilharz 1972, Kolb and Bilharz 1972). Two of the Aztec interments were found in the north central area of the courtyard and one Teotihuacan infant was interred in a subfloor ofrenda adjacent to the courtyard altar. All other human remains came from an extensive midden located outside of the south wall of a storeroom (bodega).

Non-human animal bones with butchering marks and deterioration from cooking came almost exclusively from the midden. These remains included fewer genera than TC-8:1-2 but deer, dog, cottontail, hare, and turkey were represented. Some fragmented marine mollusk shells (n = 110) were scattered throughout the eastern half of the TC-8:3 courtyard (Kolb 1973a). Of these, forty-five were Spondylus calcifer, six of which were found in subfloor ofrendas. One tubular shell bead of S. calcifer was also recovered. However, Room 3, the bodega in the southeast corner of TC-8:3 contained an additional 3,817 complete or large but fragmented specimens of S. calcifer (Kolb 1973a). The bodega, ofrendas, and other features of TC-8:3 will be detailed in a subsequent section of this monograph. (See Figures 24-28 and Appendices I-V.)

Structure TC-8:4

During the 1962 field season, one distinct structure, TC-8:4, was completely excavated (Sanders 1965:113-114), 1966:128; Sanders et al 1979:337; Kolb 1979a:378-379). This small, compact apartment complex measured 22.0 by 23.0 m, and had an exposed floor area of 529 m^2. A sunken central courtyard with small platform altar was surrounded by six porches, four of which were associated with four two-room apartments, while two porches served as "reception halls" at the east and west entrances to the complex. In its regular plan consisting of courtyard, apartments, lightwells (atria), etc., TC-8:4 resembled a miniature version of the Xolalpan urban Teotihuacan residence (Linne 1934:37-74). The 1963 field survey and mapping of TC-8:4 suggested that the excavated structure was conjoined to a larger mound which had debris scattered over an area of 2,595 m^2 (Kolb 1963, 1979:379). The eastern doorway and "reception hall" led across a narrow patio (?) or alley into the larger eastern section of the mound.

Fragments of polychrome (red, bluish green, and yellow) painted plaster were found in several locations in Porch 3 and Rooms 4 and 5. The room's plaster fragments were painted solid red, but the Porch 3 polychrome specimen was in situ and consisted of a mural in simple geometric design. All of these murals came from the same apartment.

No unusual artifact concentrations were found except for the sherds of a restorable "Red/Granular White" Ware Late Xolalpan phase amphora (Specimen 7615) from the southwestern corner of Room 4 behind Porch 3 (Kolb 1984b:50, 79, Figure 3). An almost identical specimen came from Linne's Xolalpan excavation in the urban center (1934:94-95, Figure 126) and other fragmentary amphorae were found at the urban residences of Atetelco, Yayahuala, and Zacuala Palacio and Patios (Sejourne 1959:170-171; 1966a:28, 172-174, Figures 154 and 155). These rare and unusual vessels are apparently confined to the dwellings of "high status" members of Teotihuacan society. Two Copoid Ware vessels, both copas (cups) were found on the stairways leading to the central courtyard (Kolb 1987a, 1987b).

Mound TC-8:4 was overlain by an Aztec chozo during the late Postclassic, and the remains of eight individuals (four adult and two immature, and two probable Aztec immature burials) were recovered primarily from the southeastern rooms of the complex (Bilharz 1972, Kolb and Bilharz 1972). The two probable Aztec immature individuals came from a midden immediately east of the eastern exterior wall of TC-8:4. The non-human animal bone (deer, dog, and turkey) came predominantly from the Central Courtyard and both "reception halls." Only food consumption rather than both food processing and consumption is a notable difference between TC-8:4 and TC-8:1-2 and 3 (Sanders et al 1979:343). Twenty-nine marine shells, all but one Spondylus calcifer, were scattered throughout the site (Kolb 1973a).

Summary and Interpretation of Xolalpan Phase TC-8

In summary, the upslope western one-third of the TC-8 site (ca 3.0 ha) consisted of a "religious precinct" including the temple pyramid, a small plaza, and three probable "civic" buildings located on the northeast, south, and west sides of the plaza. Seven additional mounds, probably representing small apartment compounds, were also associated. The downslope or eastern two-thirds of the site (ca 7.5 ha) was composed of twenty-one mounds, nine of which were inferred as probable large rather than small apartment complexes.

These mounds were arranged in three east-west tiers separated by small or large plazas. The excavated mounds (TC-8: 1-2, 3, and 4) were in the middle tier and separated by a large earthen plaza. The unexcavated residential mounds varied in size between TC-8: 1-2 and 3, the larger, and TC-8: 4, the smaller.

Sanders et al (1979: 341-342, 343-344) concluded that TC-8, as a Middle Horizon village, was inhabited by patrilocal lineages living in each compound or apartment complex, while each apartment housed a nuclear family, but the larger compounds could have been composed of lineages of extended families. A lineage is a consanguineal kin group, the members of which trace their common relationship through a specific series of remembered genealogical links (Murdock 1949: 46, 74-75). However, variations in compound size, plan and construction materials, as well as quality and quantity of artifacts, suggested variations in social rank which might reflect organizational levels within a calpulli (Monzon 1949: 31-32), or ramage. The Aztec calpulli was a social unit approximately the size of the entire TC-8 village, that is a kin group (clan or deme according to some authorities) composed of 20-400 extended families with endogamous and ambilateral tendencies (Monzon 1949: 68-69). As such, a calpulli was a corporate unilineal (patrilineal) descent group which coresided in a barrio (Fried 1957: 23-24; Driver 1969: 250-252).

In urban Teotihuacan, each apartment compound was a tlaxilacalli, a ranked, spatially localized descent group within the calpulli as at Tenochtitlan, where craft specalization followed barrio lines (Monzon 1949: 31-33, 48, 78). Teotihuacan social structure possibly emerged from undifferentiated egalitarian patrilineages or patriclans into conical clans or calpulli (1949: 77-78), therefore the barrio as a Teotihuacan spatial unit may have been the predecessor of the calpulli. Lineage craft specializations probably grew into barrio occupations in the urban center, such as ceramic manufacturing did at Teopancaxco and Tlajinga 33 (Storey and Widmer 1982).

Spence's analysis of human burials from urban Teotihuacan sites provides the evidence for postulating the tlaxilacalli as an exogamic, patrilineal, viripatrilocal, corporate kinship group rather than an endogamous and ambilateral descent group (Spence 1971: 82-83, 149-157; 1976: 144-145). Genetic traits in males indicated close biological affinities and kinship bonds associated with viripatrilocality and exogamy in barrios of three to fifteen apartment compounds. Admittedly the sample size was small, and only adults rather than immature individuals were studied.

The burials came from La Ventilla B (forty-five males and thirty-three females), Tetitla (ten males and five females), Yayahuala (six males and six females), Zacuala Palacio (six males and three females), Zacuala Patios (seventeen males and seven females), Oaxaca barrio (two males, one female, and one probable male), and N4W3 Crematory (a minimum of twenty-eight adults) (Spence 1971:43-83, 90-100, 104-120; 1976:132-244). At least ten of the genetic anomalies found among the 142 non-cremated adults at the urban center were shared by the thirty-eight Teotihuacan and Aztec individuals (Teotihuacan: four adult and five immature; Aztec: sixteen adult and thirteen immature) recovered in the TC-8 site (Bilharz 1972, Kolb and Bilharz 1972). I am not arguing for a continuity of exogamic kin groups over a thousand year period.

Sanders et al (1979:347) stated that "excavations in three houses revealed no evidence of craft specialization," and postulated that TC-8 was an "agriculturally based settlement." The processing of maguey pincas for fibers using the obsidian scrapers and irregular flakes argued for maguey cultivation and perhaps maize tillage as well. Following the rural model for the Aztec calpulli (Monzon 1949), the Maquixco population provided corvee labor and military troops, and sent tribute in the form of horticultural products to the urban center.

Given the fact that TC-8 lies only 5.1 km due west of the Pyramid of the Sun and may be a theoretical terminus of an extension of the West or Western Avenue which terminated to the east at the Great Compound and Miccaotli, the TC-8 "rural" village probably had more of a "suburban" character. The site could be a distant "Detached Barrio" of a Supra-Regional Center and, as such, would have served as a "dormitory town" inhabited by craft specialists who worked in the metropolis, and by some cultivators (Kolb 1979a:128). I have identified two Supra-Regional Center detached Barrios (TC-118 and 124) and twelve Barrios (TC-11, 12, 13, 14, 20, 31, 32, 116, 117, 119, 120, 121) associated with the Supra-Regional Center (TC-1).

The status differences in architecture seen at TC-8, may indicate that a possible village "chief" or other "high status" individual resided in the "miniature Xolalpan" TC-8:4 complex, and that lesser status persons residing in "Tlamimilolpa-like" TC-8:1-2 and other unexcavated residences. However, normally, we might expect a political leader to live in an apartment complex adjacent to or near the civic-religious structures in the western third of the site. The lack of evidence associated with craft specializations, and the abundance of obsidian scrapers and flakes leads to a conclusion that the TC-8 corporate group consisted of rural "peasant cultivators," but who participated more fully in the activities at the Supra-Regional Center than many of their rural contemporaries.

The marine mollusks at TC-8:3 allow the further interpretation that shell working was a part or full-time craft activity, but one not seen in the artifactual record given the kinds of tools (perishable tubular bone or wooden drills; bone, wood and/or fiber polishers; fine abrasives such as sand and pumice; etc.) such artisans would and do use (Zeitlin 1978). However, none of the bodega Spondylus specimen or those fragments found in any of the excavations showed any evidence of being worked. Only the two Spondylus tubular beads were found, but if shell working was a craft activity, the methods of excavation would not have recovered the waste from ground and polished shell, which, of course, may have been collected for other purposes. At San Jose Mogote, Oaxaca, shell ornament manufacture was limited to certain houses in one section of the site, based on the presence of unfinished pieces and shell waste (Pieres-Ferreira 1978:85). The smaller Oaxacan site of Tierras Largas had no evidence of house or barrio specializations in shell working (Winter 1973:181).

One could beg the question by contending that shell working areas were in the unexcavated tlateles or in the unexposed sections of TC-8:3 or eastern portion of the conjoining TC-8:4 site. On the other hand, such activities may have been conducted in the large earthen patio between TC-8:1-2, 3 and 4, and that the artisans swept up debris and disposed of it on a daily basis as Aztec craft apprentices were required to do. The quantity of shells and their contexts provide an alternative explanation, subsequently detailed, namely that the TC-8 Santa Maria Maquixco el Bajo site residents were peripheraly, or perhaps significantly, involved in the transportation of marine shell which was being imported into the Supra-Regional Center (urban Teotihuacan, TC-1).

FIGURE 24: SANTA MARIA MAQUIXCO EL BAJO SITE EXCAVATIONS
(TC-8:1-2, 3, 4): -- EXAMPLES OF Spondylus calcifer
Carpenter, 1857 (PANAMANIAN MARINE FAUNAL PROVINCE)

(photograph by Kolb)

(left to right)

Small unworked specimens: 8552, 8552, 8672
Medium-size unworked specimen: 8640
Tubular bead (center): 8114

FIGURE 25: SANTA MARIA MAQUIXCO EL BAJO SITE EXCAVATIOS (TC-8: 1-2, 3, 4): VARIOUS UNIDENTIFIED, WORKED MARINE SHELL SPECIMENS

(photograph by Kolb)

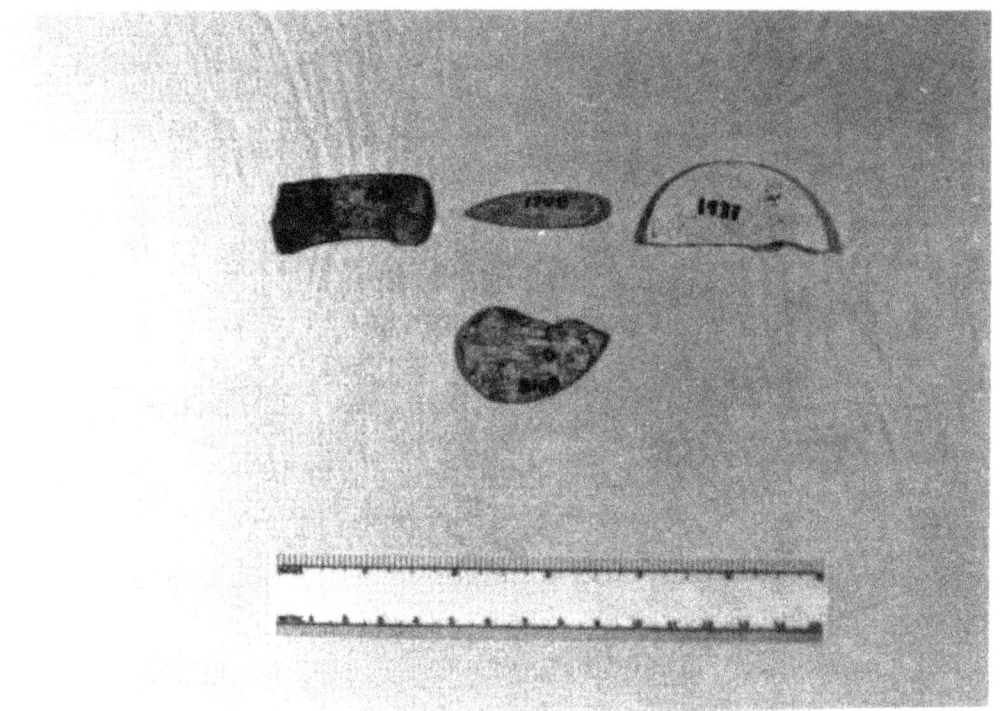

(Top Row, left to right)

Cut and abraided specimen: 1561
Cut and abraided specimen, conical-shaped: 1708
Cut and abraided specimen, disk-shaped pendant (?): 1737

(Bottom Row)

Cut, abraided and drilled specimen, disk-shaped pendant (?): 8169

FIGURE 26: SANTA MARIA MAQUIXCO EL BAJO SITE EXCAVATIONS
(TC-8:1-2, 3, 4): TWO IDENTIFIED, UNWORKED MARINE SHELL
   SPECIMENS (PANAMANIAN MARINE FAUNAL PROVINCE)

(photograph by Kolb)

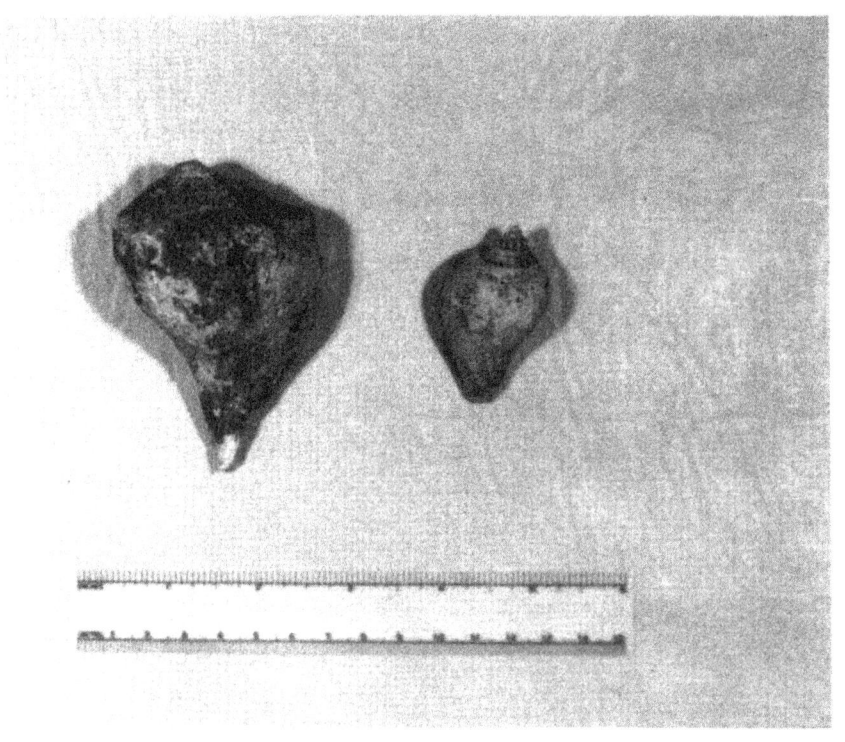

(left to right)

Fasciolaria princeps Sowerby, 1825 OR F. gigantea Kiener,
   1840: 7631
Melongena (prob.) patula Broderip and Sowerby, 1829:
   8820

FIGURE 27: SANTA MARIA MAQUIXCO EL BAJO SITE EXCAVATIONS
(TC-8:1-2, 3, 4): THREE IDENTIFIED, UNWORKED MARINE
SHELL SPECIMENS (PANAMANIAN MARINE FAUNAL PROVINCE)

(photograph by Kolb)

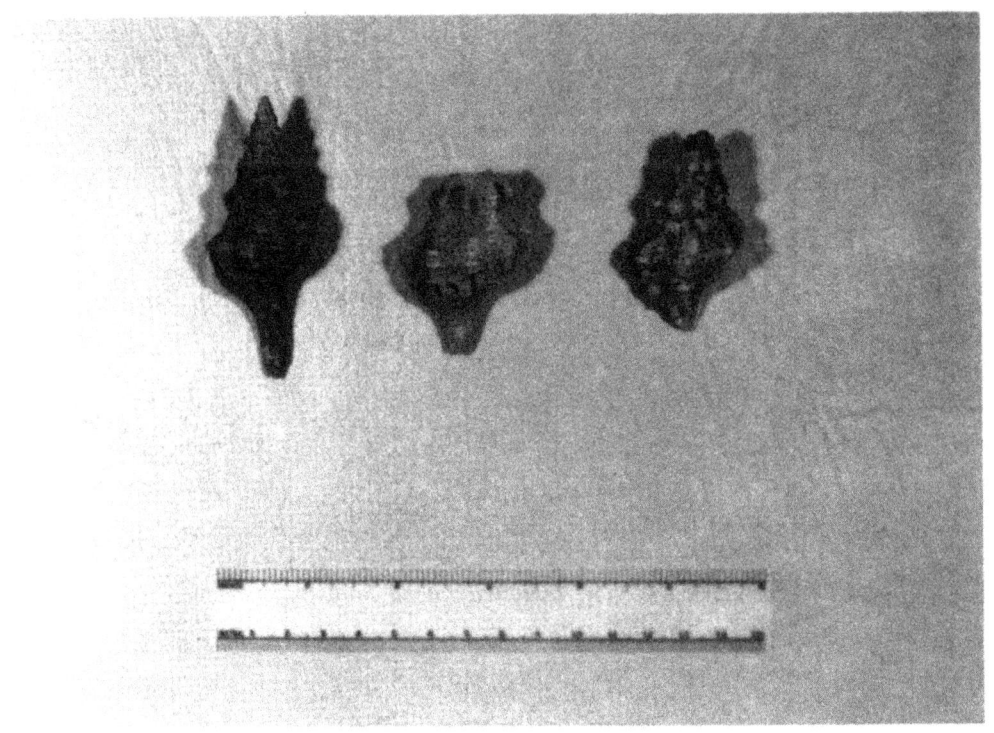

(left to right)

<u>Latirus ceratus</u> Wood, 1928: 8517
<u>Latirus ceratus</u> Wood, 1928: 8537
<u>Latirus ceratus</u> Wood, 1928: 8809

FIGURE 28: SANTA MARIA MAQUIXCO EL BAJO SITE EXCAVATIONS
(TC-8:1-2, 3, 4): Unionidae (Family), Elliptio (Genus)
"FRESHWATER CLAM" WORKED SPECIMENS

(photograph by Kolb)

(Top Row, left to right)

Worked specimens, disk-shaped, beads or pendants (?):
    1760, 8138, 8147, 8147, 8164, 8168

(Bottom Row, left to right)

Worked specimens: 8200, 8553, 8555

CHAPTER SIX

THE SHELLS OF MAQUIXCO AND OTHER "RURAL" SITES

Introduction

Sanders (1966:140, Sanders et al 1979:353) noted the concentration of Spondylus calcifer shells at Santa Maria Maquixco el Bajo but erroneously associated them with the midden at TC-8:3 rather than Room 2 of that structure (Kolb 1973a, 1979a:223-224). There was meager evidence of shell working at the TC-8 site; one purplish red Spondylus calcifer bead was recovered in the eastern section of the TC-8:3 central courtyard and a second in TC-8:1-2, while a Unio (freshwater clam) artifact was associated with subfloor ofrenda in the former residence. It is unlikely that fresh shellfish in their shells were imported from the Pacific Coast (Panamanian Marine Faunal Province) as a food resource. The Spondylus and other genera of shell were "clearly not brought as raw material for artifacts" according to Sanders (1966:140, Sanders et al 1979:353), but I believe Sanders' supposition to be incorrect. The TC-8:3 Room 2 bodega is unique. To my knowledge no other site in the Meseta Central has marine shell in this abundance, and certainly not localized an an interior room. (See Figures 23-27 and Appendices I-V.)

TC-8:3 Platform Altar Ofrendas

The Central Courtyard platform altar in TC-8:3 was rectilinear and had a talud and tablero similar to those in TC-8:1-2 and TC-8:4, as well as altars in urban center residences such as Xolalpan (Linne 1934:37-49), Yayahuala, and Zacuala Palacio (Sejourne 1966b:192-201). The TC-8 altar had maximum dimensions of 3.26 m east-west and 3.00 m north-south inclusive of apron, and 2.30 m east-west and 2.57 m north-south exclusive of the 18.0 cm apron (8.79 m^2). Maximum preserved height was 30.0 cm. No concentrations of ceramics, censers, figurines or other artifacts were associated in, on, or near the platform.

Approximately twenty percent of the original courtyard floor was preserved, including the area surrounding the platform altar. A one meter wide trench was excavated through the floor along the east side of the altar in order to examine the phases of construction.

The floor had been constructed in a series of levels following the clearing of soils to expose the natural tepetate surface. The courtyard floor was leveled by cutting away portions of the natural surface and filling depressions with coarse, crushed tepetate. Over this base a thin layer of fine crushed tepetate was added and packed or tamped down prior to the application of a "stucco" subfloor made from a mixture of crushed volcanic scoria (tezontli), earth, and a lime cement. A thin coating of white lime plaster completed the unpainted floor. In some urban residences, floors may have been painted with a specular red hematite pigment.

The platform altar apron was covered with the same white lime plaster as the floor, indicating that they had been plastered at the same time. Two separate floor sequences (plaster, "stucco," and finely crushed tepetate) were found in the altar area, but closer examination revealed that the platform altar and first courtyard floor were contemporary constructions, probably dating from the initial phase of residence construction. Below both floors at the extreme northeast corner of the platform, an 18.0 cm deep, 34.0 cm diameter semi-conical pit had been excavated in the natural tepetate surface, and contained an altar ofrenda. The north edge of this subfloor pit was 22.0 cm south of the northeast corner of the altar. The one-meter wide trench through the floors was continued around the entire altar, and a second subfloor ofrenda was found on the north side near the northeast corner and was apparently contemporary with the first offering (Kolb 1962:August 13-14).

The northeast corner ofrenda on the east side of the platform contained two ceramic vessels, both San Miguel burnished tan carinated basins, one smaller inverted at a 25° angle to the southwest resting inside the larger. The smaller basin was complete and unbroken and had a lattice-like double-line incised design forming two "diamonds" with incised circles within the double lines at the apices. Where the diamond patterns met, the circles had been painted red. The larger vessel was fragmented (n = 30) but restorable, save a totally missing 2.5 cm long rim sherd. This basin was decorated with a series of incised outlines of birds (eagles?) in profile with outstretched wings. Where wingtip stretched toward wingtip, an incised red-painted circle conjoined the series.

A layer of fine "sand" filled half of the larger basin, and the rim of the smaller vessel rested on this layer. Microscopic examination indicated that this sandy material was composed of well-rounded particles and would have been derived from any fluvial context such as a lacustrine or marine (ocean) beach. The grains consisted of quartzite, quartz, alkalai feldspars, plagioclase feldspar, basaltic hornblende, hornblende, and epidote. This array suggested the "sand" had a probable Basin of Mexico origin. Beneath this layer and on the bottom of the basin were three biconically drilled "greenstone" beads (jadeite or serpentine), four small, complete Spondylus calcifer Carpenter, 1857 bivalves, and three small, badly deteriorated bones. These tubular bones could not be specifically identified, but, since two were tubular without marrow cavities characteristic of mammalian species, they were most likely from a bird (Aves). Neither appeared to have been tubular bone drills. One undiagnostic Alpha "Thin Orange" Ware sherd from a hemispherical bowl was found beneath the second floor adjacent to the northeast corner of the platform altar (Kolb 1965b, 1973c, 1977, 1982, 1984b).

The second ofrenda, contained within a subfloor pit (semicircular-half conical 30.0 cm wide, 65.0-70.0 cm across, and up to 25.0 cm deep) dug into tepetate, was located on the north side of the altar, 18.0 cm west of the northeast corner (Kolb 1962:August 14). Within this pit, also sealed by the double floor previously described, a badly deteriorated and fragmented burial was recovered. The remains (seven teeth, six cranial fragments, two vertebrae, four rib fragments, and four segments of long bone diaphyses) were those of a human infant, aged at about one year at time of death. The distribution of the skeletal fragments suggested a secondary, non-bundle burial (Bilharz 1972, Kolb and Bilharz 1972).

The earth-filled pit also contained two small, complete Spondylus calcifer Carpenter, 1857 shells and one thin, flat freshwater clam, Uno discus Say, 1838 shell modified into a trilobed artifact with two perforations (Specimen 8321). I originally thought this was a pendant, but now believe it to be a representation of a bird's wing, possibly a part of an articulated, "puppet-like" ornament. Pires-Ferreira (1978:85) called a similar artifact from San Jose Mogote, Oaxaca, a pendant with a "paw-wing" motif. Immediately below the two floors and above the burial were three undiagnostic Alpha "Thin Orange" Ware sherds, one of which fit the fragment associated with the nearby ceramic ofrenda.

The "Thin Orange" sherds suggest that these ofrendas were contemporary, while altar building techniques confirmed that both subfloors offerings were made at the time of the construction of the platform altar and first floor above tepetate. The two incised, red-painted carinated basins were dated to the Early and Late Tlamimilolpa phases, ca 300-500 A.D., so that the initial floor and altar dated to this period. The analysis of sherds found on the adjacent floor of the courtyard dated from Early Tlamimilolpa through Late Xolalpan phases (ca 300-650 A.D.) Therefore, at least part of the TC-8:3 structure construction dated to Early Tlamimilolpa and subsequent phases.

Both the subfloor ofrendas, the first with carinated basins and the second, the infant interment, had small, complete specimens of Spondylus calcifer, which has as its habitat the Panamanian Marine Faunal Province. The Unio artifact is of possibly "local" material, but the "greenstone" beads also point to external Basin of Mexico sources. No other ofrendas, features, or burials at TC-8:1-2, 3, and 4 contained marine or freshwater shells or artifacts. As previously noted, there were thirty-nine fragments of Spondylus calcifer scattered throughout the eastern half of the TC-8:3 courtyard, plus the tubular shell bead.

TC-8:3 Room 2 Bodega

Room 2 at TC-8:3, which I have characterized as a bodega (storeroom), comprised 22.08 square meters of interior floor space (Kolb 1962, 1973a). The walls of the room were built of rough-cut stone (tezontli, tepetate, and other local rock) and, unlike the other rooms and walls, never had "stucco" or plaster finish, nor was any floor constructed above the natural tepetate surface (cf. Flannery 1974a:19-23). The height of the preserved walls varied from 50.0-150.0 cm, and the room interior was excavated in fifty-four arbitrary stratigraphic squares and levels. In the lower levels, and continuing to the tepetate surface, were 3,817 specimens -- complete and fragmentary -- of Spondylus calcifer Carpenter, 1857. These specimens varied from shells 15.0 cm across to fragments less than 3.0 cm in diameter. Literally, there was more shell than earth in several of the excavated squares (Kolb 1973a).

Approximately 4,100 sherds (no complete or identifiable fragmented vessels) were recovered in levels in or immediately adjacent to the room. The uppermost levels, which were within the plow zone, had a mixture of predominantly Aztec sherds but included some Classic Teotihuacan sherds as well. The levels immediately above the tepetate and those in which over ninety percent of the shells were found contained almost exclusively Classic ceramics. Less than four percent of the 4,100 sherds were Aztec, while the Classic pottery and ceramic figurines all dated to the Early and Late Xolalpan phases (ca 500-650 A.D.).

The few Aztec sherds in association with the shell can be accounted for by rodent activity (tusas or groundhogs inhabited the TC-8 village area) and/or the planting of maguey (Agave americana) possibly during the Aztec occupation of the site, and certainly during the Colonial era when Hacienda de la Cadena after 1609 used the site as a maguey plantation. I therefore concluded that Room 2 at TC-8:3 served as a storeroom for Pacific coast shells being imported into Teotihuacan during the Xolalpan phases and, to judge from the ofrendas, probably as early as the Early Tlamimilolpa phase, e.g. potentially from ca 300 A.D. to at least 650 A.D. if not later.

TC-8:3 "Foreign" Connections

In the midden deposit immediately south of Room 2, the fragments of a carved tepetate Huehueteotl brasier (Specimen 8150) and two sherds attributed to the Classic Lowland Maya were recovered (Kolb 1962, 1972a). The sherds were examined by R. E. W. Adams (1972), T. Patrick Culbert (1972), the late James Gifford (1971, 1972), Robert Rands (1972), and Robert Sonin (1972). They unanimously agreed to an Early Classic Lowland Maya origin, most likely the Tikal Manik phase, while Culbert specified one sherd as "Mojara Orange Polychrome, Mojara Variety," reported from Altar de Sacrificios (Adams 1971:36, 126-127). The chronology would be 300-600 A.D. Gifford (1972) saw "direct comparison counterparts at Barton Ramie," Belize. Associated in the same context as these sherds was a human frontal cranial fragment which had evidence of frontal (or fronto-occipital) cranial deformation, a characteristic of the Lowland Classic Maya (Bilharz 1972, Kolb and Bilharz 1972).

Millon (1964:351; 1967a:45; 1973:28, 34, 42; 1981:226) has reported Classic Lowland Maya ceramics, including Tzakol sherds, from the eastern and northwestern edges of the Supra-Regional Center, while Linne (1934:96-99, 186; 1942:98, 178, Figures 328-329, 331, Pl. 2) noted "Maya style" sherds in the Xolalpan residence, and recovered "Peten Maya" (Uaxactan II and Holmul II-IV) ceramics from a sealed deposit at the Tlamimilolpa site. Classic Teotihuacan ceramics were found at Tikal (W. Coe 1972:258, 268), hence we should expect to find the reverse.

Immediately to the north of Room 2 in TC-8:3, at the junction of Alley 1 and Patio 1, a unique, non-Teotihuacan rim sherd was found in the excavation level next to the preserved floor. Phil Weigand (1972) stated that it "... appears to be from the Chalchihuites area during the Classic" Teotihuacan Period, ca 250-500 A.D. Five other sherds having origins in West Mexico -- Jalisco, Nayarit and Zacatecas -- also dated to the Classic, and one other Classic Lowland Maya sherd came from the TC-8:1-2 and 3 excavations (Weigand 1972, Kolb 1972a). In addition, twelve Monte Alban III-A and III-B gray sherds (possibly locally made in Millon's Oaxaca Barrio), and eight Tajin Huastec III sherds and other Gulf Coast sherds of the Classic and Postclassic were recovered at TC-8 (Ekholm 1970, Kolb 1972a). Similar Totonacan, Huastecan, and Oaxacan pottery was also found at the urban Tlamimilolpa site (Linne 1942:178, Figures 322-326).

Preliminary Hypothesis

The few sherds attributed to the Classic Lowland Maya (Altar de Sacrificos, Barton Ramie, Tikal Manik-Tzakol), to West Mexico (Chalchihuites, Zacatecas, Jalisco, Nayarit), to the Gulf Coast (Huasteca), and potentially the Valley of Oaxaca (Monte Alban), as well as the frontally deformed (Maya?) human cranial fragment do not necessarily argue for direct culture contact of the TC-8 site inhabitants with these "foreign" regions. Similarly, the quantities of Alpha "Thin Orange" ceramics found at TC-8 and other rural Classic sites did not argue for direct contact (Kolb 1973c, 1977, 1984a, 1986). The potential chronology, based on the ofrendas, bodega, and "foreign" sherds, ranged from ca 250-650 A.D. The sherds and cranial fragment might have been "souvenirs" from the journeys of others to these regions or could have been collected by TC-8 residents, especially children, from middens in the urban center, potentially in market areas where foreign goods would be concentrated.

The quantities of Spondylus calcifer shell brought from the Pacific Coast do argue for more direct contact by TC-8:3 residents with peoples from West Mexico. It is important to recall that the Maquixco site is only 5.1 km due west of the Great Compound, and on a line with a prolongation of the Western or West Avenue, hence "en route" toward the Pacific Coast. There is no evidence suggesting that any shell working was being done at TC-8, but TC-8:3 residents were likely involved in the shell procurement and/or distribution. The Room 2 concentration argues for a bodega from which unworked shell was transported to the workshops of the Supra-Regional Center for conversion into ornaments (beads, bracelets, pectorals, etc.) or for use in ritual and/or ceremonial activities (Kolb 1973a, 1979a:223-224).

Millon reported that a "group of lapidaries" in a barrio in Square N3E5 (TE-18), two kilometers east of the Pyramid of the Sun, specialized in working jade (and related "greenstone"), onyx, and the shell of only *Isognomon alatus* Gmelin, 1791, the "Flat Tree oyster" (Starbuck 1975:124, 150; Millon 1981:227). This is a bivalve common in mangrove swamps in the Caribbean Marine Faunal Province (Warmke and Abbott 1962:165). Shell working in *Chama echinata* Broderip, 1835, a Panamanian species ranging from the Gulf of California to Panama, was confined to the Oaxaca *barrio* (Keen 1971:147, Starbuck 1975:150-151, Millon 1981:227). The possible working of *Spondylus* took place in the urban center in the "Old City," Oztoyahualco, in Square N6W3 northwest of the Pyramid of the Moon (Millon 1973:38-39, 1981:227). Unfortunately, we know little of the nature of this possible workshop. Tlajinga 33 (Storey and Widmer 1982:39) had craft specializations in shell and in fine-grained travertine during the Early and Late Tlamimilolpa phases. The specific evidence for shell artifact production was in the form of shell debitage. Further elaborations are needed. Zeitlin (1978:183-210) has described the Laguna Zope shellworking center on the southern Isthmus of Tehuantepec. I know of no other adequately described shell artisan sites.

Residents from TC-8:3 could have been the suppliers of the *Spondylus* workers at Oztoyahualco, as these sites are also only 5.0 km apart, and Tlajinga 33 is only about 6.5 km distant. The Barranca de Malinalco presently is a physical barrier between the Maquixco and Oztoyahualco sites, but in Classic times the barranca may have been nonexistent or insignificant, but if a barrier, may have been bridged by a wood and/or rope walkway.

While it seems likely that the TC-8:3 inhabitants were short-distance suppliers, perhaps the tumpline porters (*tlamemes*) or their employer-supervisors, the residents might merely have provided the storage facility (*bodega*) for *pochteca* long-distance merchants (Acosta Saignes 1945; Bittman and Sullivan 1978; Sahagun 1959:1-32; Kolb 1984a:218, 1986:191-192). These *pochteca* could have been "middlemen" or "wholesalers" in the procurement and distribution of *Spondylus* to the shell workshop at Oztoyahualco and Tlajinga 33 and/or consumers at the Great Compound, Teotihuacan's largest market. However, such *pochteca* could have been the direct procurers of shell from the coast, and potentially included "shell divers" among their ranks, although this seems unlikely.

A Summary of Teotihuacan Valley Rural Site Mollusks

Seventeen genera and/or species of marine and riverine (freshwater) mollusks were identified from excavations and the surface reconnaissance of rural Classic sites in the Teotihuacan Valley from 1960-1965 (Sanders 1965, Sanders et al 1970). These specimens included ten Gastropoda (four species from the Panamanian Marine Faunal Province, four species from the Caribbean Marine Faunal Province and two genera from either/or the Panamanian or Caribbean Provinces), and seven Pelecypoda (four species from the Panamanian Marine Faunal Province) and three species which have riverine habitats. Mollusks were recovered from only eight of 134 Classic sites (TC-2, 8, 10, 13, 40, 49, 73, 91), including three excavated or tested sites (TC-8, 10, 49). Specimens were recovered from Classic sites in the following ecological zones of the Teotihuacan Valley (Sanders 1965): Delta (TC-10), Lower Valley (TC-2, 8, 13), Upper Valley (TC-91), and North Slope of Cerro Gordo (TC-40, 49, 73). These data are summarized in Appendices I-V. (See Figures 29 and 30.)

A total of 4,200 worked and unworked mollusk shells and fragments composed the analytical sample. The identification of marine and riverine specimens was initially undertaken by the author in 1962, and was continued by Lawrence H. Feldman (1968a), who classified approximately fifteen percent of the samples. The unstudied materials and those previously classified were also examined by Harold S. Feinberg, Assistant Curator, Department of Living Invertebrates, American Museum of Natural History (1971, 1972, 1973) and the author. Feinberg gave valuable assistance on identification procedures, data reliability, and problems of classification taxonomy. His assistance is gratefully acknowledged, especially for his clarification of Spondylus calcifer Carpenter, 1857, Spondylus princeps Broderip, 1833 (both Panamanian Marine Faunal Province species), and Spondylus americanus Gmelin, 1791 (the Caribbean Marine Faunal Province species).

Of the 4,200 specimens from Classic sites, only eight were from surface reconnaissance collections, while the remaining 4,192 specimens came from the four excavated sites and had the following distributions: Site TC-8 (Santa Maria Maquixco el Bajo), mounds TC-8:1-2, 3 and 4, a total of 4,188 specimens (Classes: Gastropoda eleven, Pelecypoda 4,148, indeterminate twenty-nine; Ranges: Caribbean M. F. P. five, Panamanian M. F. P. 4,138, Caribbean or Panamanian M. F. P. thirty, Fresh Water [Riverine] fifteen). Site TC-10:2 (Venta de Carpio), a total of three specimens (Classes: Gastropoda two, indeterminate one; Ranges: Caribbean M. F. P. one, Caribbean or Panamanian M. F. P. two). Site TC-49:1-3 (Tenango), a single specimen (Class Pelecypoda, Range Fresh Water [Riverine]).

FIGURE 29: VENTA DE CARPIO SITE EXCAVATION (TC-10:2):
THREE UNWORKED MARINE SHELL SPECIMENS

(photograph by Kolb)

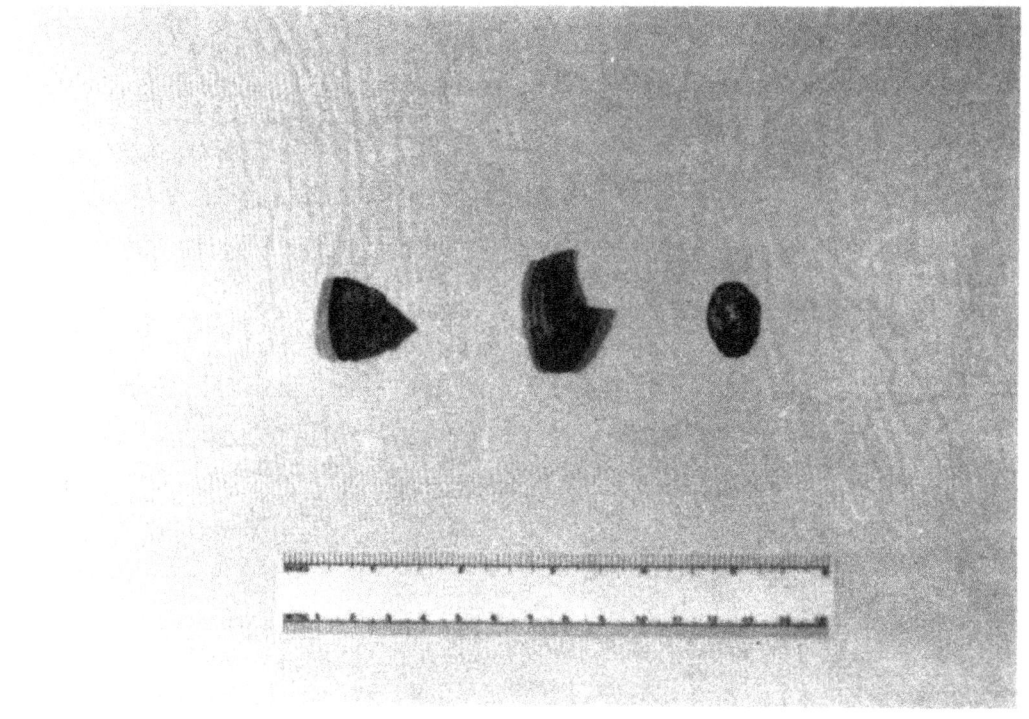

(left to right)

Unidentified, burned fragment: 10695
Oliva sayana Ravenel, 1834 (Caribbean) OR Oliva porphyria
    Linnaeus, 1758 (Panamanian), burned fragment: 10761
Fissurella angusta Gmelin, 1791 (Caribbean), unworked and
    burned: 10713

FIGURE 30: TEOTIHUACAN VALLEY PROJECT CLASSIC PERIOD
SURFACE SURVEY (TC-2, 8, 13, 40, 73, 91): VARIOUS
IDENTIFIED AND UNIDENTIFIED MARINE SHELL SPECIMENS

(photograph by Kolb)

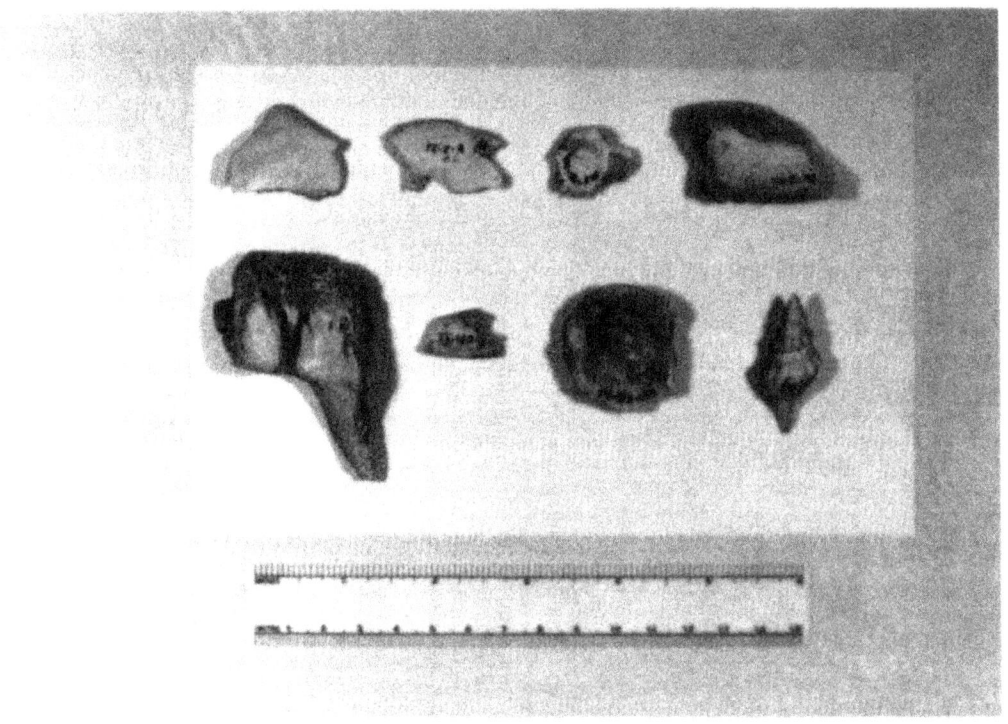

(Top Row, left to right)

Cut and abraided fragment, possible cup, ornament or atl
    atl finger loop: TC-2:1 (1)
Cut, abraided and drilled fragment, possible bead or
    pendant: TC-2:1 (2)
Cut fragment, probably Spondylus calcifer Carpenter, 1857
    (Panamanian): TC-8:14
Cut and abraided fragment, Glycymeris subabsolata
    Carpenter, ... (Panamanian): TC-8:28

(Bottom Row, left to right)

Unworked fragment, Chama buddiana C. B. Adams, 1858
    (Panamanian): TC-13:4 (D)
Unworked fragment, unidentified: TC-40:7
Unworked specimen, Chama echinata Broderip, 1835
    (Panamanian): TC-73:36
Unworked specimen, Glycymeris subabsolata Carpenter, ...
    (Panamanian): TC-91:1(A)

Site TC-8 had the following distributions: TC-8:1-2 excavations produced 215 specimens (Class: <u>Gastropoda</u> six, <u>Pelecypoda</u> 202, indeterminate seven; Range: Caribbean <u>or</u> Panamanian M. F. P. eight, Panamanian M. F. P. 199, Fresh Water Riverine eight). Mound TC-8:3 yielded 3,927 specimens (Class: <u>Gastropoda</u> four, <u>Pelecypoda</u> 3,902, indeterminate seven; Range: Caribbean M. F. P. four, Panamanian M. F. P. 3,895, Caribbean <u>or</u> Panamanian M. F. P. twenty-one, Fresh Water [Rive<u>rine</u>] seven). Mound TC-8:4 produced twenty-nine specimens (Class: <u>Gastropoda</u> one, <u>Pelecypoda</u> twenty-eight; Range: Caribbean M. F. P. one, Panamanian M. F. P. twenty-eight). No specimens were recovered from the TC-8:Pyramid. Seventeen additional shell specimens had been excavated in 1961-1962, but had lost their provenience designations (Class: <u>Pelecypoda</u> sixteen, indeterminate one; Range: Panamanian M. F. P. sixteen, Caribbean <u>or</u> Panamanian M. F. P. one).

Therefore, of the 4,188 specimens excavated at the site, one species, <u>Spondylus calcifer</u> Carpenter, 1857, comprising 4,084 specimens (97.52%), dominated the excavated sample. <u>Spondylus calcifer</u> Carpenter, 1857, was represented as follows: TC-8:1-2, 194 of 215 specimens; TC-8:3, 3,862 of 3,927 specimens; and TC-8:4, 28 of 29 specimens. Of the 3,862 whole or fragmented shells from TC-8:3, 3,817 were found in or immediately adjacent to Room 2, while the remaining forty-five were scattered throughout the eastern half of the courtyard or other architectural units or were in the <u>ofrendas</u> previously described.

The Santa Maria Maquixco el Bajo (TC-8) site specimens were definitely <u>Spondylus calcifer</u>, and were nearly exclusively adult specimens. It is possible, but not likely, that divers ventured to depths of ten meters or more to obtain the live mollusks, since they are, in their native habitat, firmly attached to substrata, requiring major efforts to dislodge them even with metal tools let alone with a "neolithic" tool kit. Shells broken from the substrata by storms would be deposited on beaches, and were probably collected from this locale. Dead <u>Spondylus</u> would also be cast ashore, but would have lost their prized "feathery" fronds.

However, in Postclassic Aztec times skilled divers were employed in the Kingdom of Colima, where an annual tribute of 1,600 <u>Spondylus</u> valves was paid to Moctezuma, as recorded in the <u>Libro de Tributos</u> (Peñafiel 1890:83, Pl. 16; Boekelman 1935:262-264). Feldman (1973) holds the opinion that the "Colima" referred to was the coastal town of Colima, Michoacan rather than the kingdom of the same name. Something along the lines of <u>Dentalia</u> "shell fishing" from open boats with elaborate gear, as was practiced by the Nootka of the Northwest Coast of North America (Drucker 1965:151-152), is suggested. Spondylus princeps was an important barter item on the west coast of Mesoamerica, as reported by Gonzalo Fernandez de Oviedo y Valdes in Historia general y natural de los Indios (1855 [4]:122). These were the Conchas coloradas known in Postclassic times into the period of Hispanic contact (Linne 1942:151).

Other Postclassic parallels can be found in the Codex Mendocino and Matriculata de Tributos (Clark 1938, Barlow 1949). The coastal Province of Cihuatlan on the Tarascan frontier was required to provide annually, among other goods, eight hundred red seashells ("like scallops") to the Aztec/Culhua Mexica. Although the province is poorly documented, the tribute towns included eighteen communities, three of which (Acapulco, Punta de Iztapa, and Bahia de Cihuatanejo) were on the coast, while Cacatulan was located at the mouth of the Rio Balsas (Barlow 1949:8-15). The inland Province of Ocuilan, today the region of Toluca west of the Basin of Mexico, paid a semiannual tribute of four hundred bundles of little cotton mantles decorated with seashells (1949:25-26). The province was located slightly east of the Rio Balsas drainage.

It would appear that the red seashells were Pecten, while the shells decorating the mantles were possibly cut and drilled Oliva or Olivella "tinklers." The Ocuilan peoples must have obtained their shells from Cihuatlan or possibly other peoples not affiliated with the Aztec "Empire."

CHAPTER SEVEN

MAJOR MOLLUSK DISTRIBUTIONS

Introduction

Several regional syntheses characterize the mollusks recovered from archaeological sites in Mesoamerica. The two for the Maya region were by Andrews (1969), who tabulated 192 species, and by Feldman (1972), with 221 species listed. From the Chiapas region, Chavez (1969) detailed 38 species, while the Valley of Oaxaca synthesis by Pires-Ferreira (1970) included 48 species for the Early and 44 species for the Late Preclassic. MacNeish et al (1967:147-149) identified only two specific genera (Marginella and Oliva) among Tehuacan Valley shell artifacts. Feldman (1974) tabulated the species found in West Mexico by area: Nayaritan (57 species), Ahuacatlan (137), Purificacion (6), Coliman (67), Rio Balsas (43), Chapalan (55), and Chichimecan (16). An initial tabulation of archaeomollusks in sites located in the Meseta Central, West Mexico, Gulf Coast, Yucatan Lowlands, and Guatemalan Highlands by the author included 192 species (Kolb 1973a). Sixty-three species were recorded for Classic sites in the Basin of Mexico, see Appendix IV.

In the current presentation, I shall consider only the species within three genera: Fasciolaria, a gastropod with five species; Strombus, a gastropod having nine species; and Spondylus, a pelecypod with seven species. Not all species of the Gastropoda conchs and Pelecypoda "thorny oyster" bivalves were found in Basin of Mexico sites associated with any one period -- Preclassic/Formative, Classic, and Postclassic. Only the Classic will be detailed here. Relevant occurrences of the species of these genera in the Maya, Chiapan, Oaxacan, and West Mexican regions (Preclassic through Postclassic) will be noted. No specimens of Fasciolaria, Strombus, or Spondylus were reported for Sonora (Drake 1960, Gifford 1946:217), nor were these mollusks represented in the early excavations at Chametla, Sinaloa or Autlan, Jalisco (Kelly 1938:4, 1945a:72).

Spondilidae

Specimens of Spondylus were reported from Oztoyahualco (the "Old City"), the Palacio de Quetzalpapalotl, and the residences of Tetitla, Yayahuala, and Zacuala at Teotihuacan during the Classic

Period. These Spondilidae (species unspecified) may have come from either or both marine faunal provinces. Spondylus americanus, the Caribbean species, was found only in the Tlamimilolpa residence at Teotihuacan. Panamanian Spondylus calcifer specimens came from the Temple of Quetzalcoatl in the Ciudadela, the surface of the nearby Great Compound, Tlajinga 33 and the Maquixco site (TC-8: 1-2, 3, 4). Spondylus princeps, also from the Panamanian province, were recovered at the Tlamimilolpa and Xolalpan residences. No other Classic Basin of Mexico sites reported Spondylus in their artifact collections. (See Figure 31, Spondilidae Distributions in Mesoamerica.)

Spondilidae (unidentified species) were found at Kaminaljuyu, Tikal, and Zaculeu in the Maya region (Feldman 1972:125, 132-133; Kidder et al 1946:145-147; Moholy-Nagy 1963:68, 70-71; Woodbury and Trik 1953:268-274). At Tikal the shell was fashioned into spangles, mosaics, and human silhouette figurines. Preclassic sites in the Valley of Oaxaca also had Spondylus shell (Pires-Ferreira 1978:89). In West Mexico, Spondylus (unidentified species) was recovered from Once Pueblos and Playa del Tesoro, both in Colima, and Cojumatlan and San Gregorio in Michoacan, as well as San Blas, Nayarit (Feldman 1968: 170, 1973, 1974:232-233; Lister 1949).

The Panamanian species Spondylus calcifer was noted at Chiapa de Corzo (Chavez 1969:219). It was not reported from the Guatemalan Highlands, Yucatan Lowlands, or Tehuacan Valley, but S. calcifer was excavated from the Oaxacan sites of San Jose Mogote and Laguna Zope (Pires-Ferreira 1978:89). Definite specimens were found at Yurecuaro, Michoacan in West Mexico, while probable specimens were thought to come from Amapa, Nayarit, Cojumatlan, Michoacan, and Playa del Tesoro, Colima (Feldman 1968b:170, 1973, 1974:231, 235; Lister 1949).

Spondylus princeps, inclusive of the subspecies princeps (S.p.p.) and unicolor (S.p.u.) specimens, were frequently represented in the Maya region. The species was noted at Copan, Piedras Negras, Pusilha, San Jose, Tikal, Uaxactun, and in Guatemalan Highland sites at Lake Amatitlan (Andrews 1969:39, 42, 43, 49, 52; Borhegyi 1966a:360; W. Coe 1959:462-468; Kidder 1947; Longyear 1952; Moholy-Nagy 1963:67; Ricketson and Ricketson 1937; Smith et al 1950:92; Stromsvik 1942:63-96; Thompson 1939). S. princeps was reported at Chiapa de Corzo (Chavez 1969:219), and S. pictoram (princeps) came from Tierras Largas, Oaxaca (Pires-Ferreira 1978:89). In West Mexico S. princeps was found at San Gregorio and Yurecuaro, both in Michoacan (Feldman 1974:235, Goggin

FIGURE 31: <u>Spondilidae</u> DISTRIBUTIONS IN MESOAMERICA

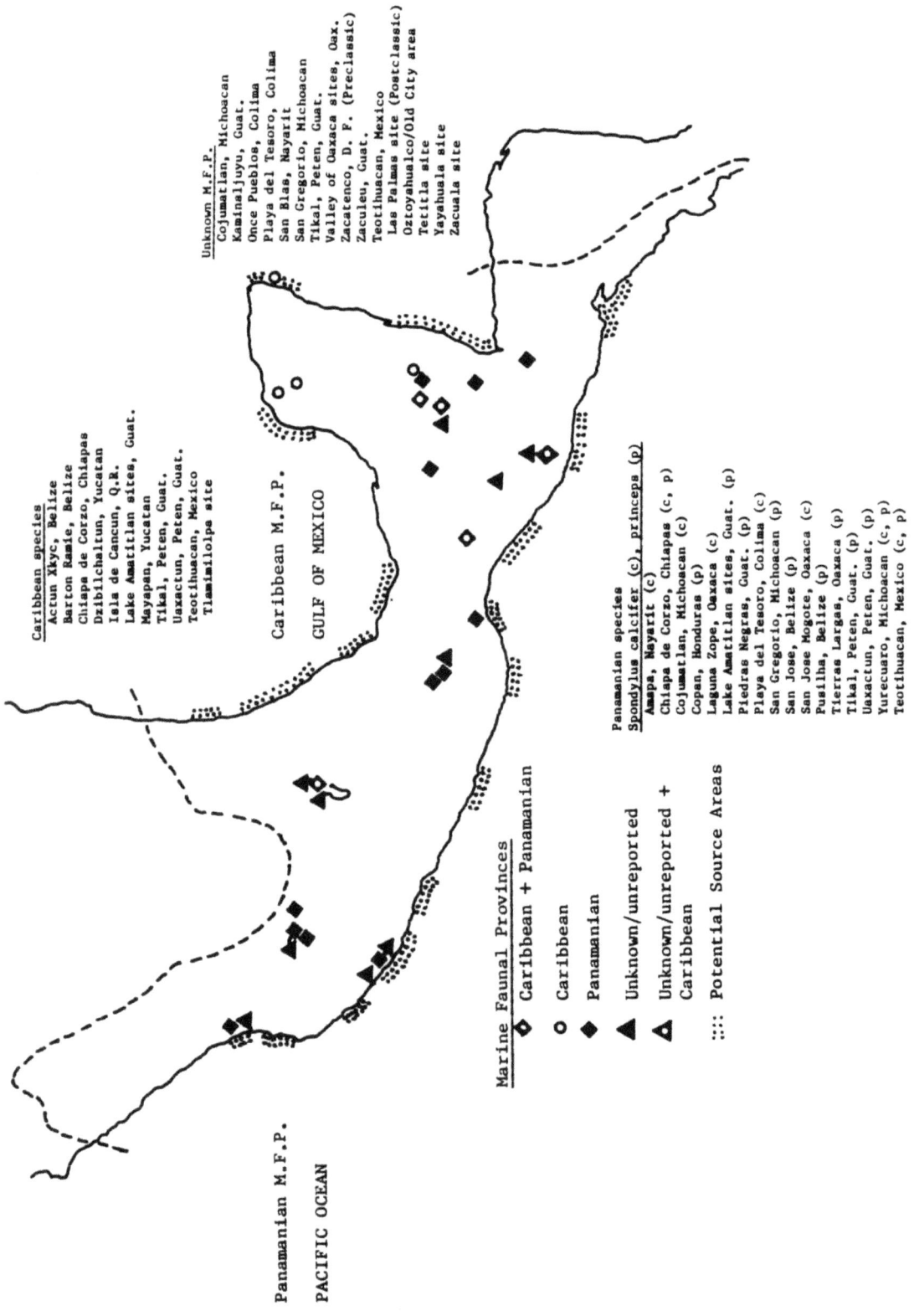

1943). No examples of the Panamanian species S. ursipes are known from Mesoamerican sites.

Spondylus americanus, the Caribbean species was especially well-represented in Maya sites, including Actun Xkyc, Barton Ramie, Dzibilchaltun, Isla de Cancun, Mayapan, Tikal, Uaxactun, and Lake Amatitlan sites (Andrews 1969: 39, 45, 47, 50, 57, 58; Boekelman 1935: 262-264; Borhegyi 1966a: 360; Moholy-Nagy 1963: 67; Proskouriakoff 1962: 321-446; Smith et al 1961). The S. americanus was represented at Chiapa de Corzo (Chavez 1969: 219) but, apparently no further north or west, e.g. not in the Valley of Oaxaca, Tehuacan Valley, or West Mexico. It is possible the specimens from Tlamimilolpa, Teotihuacan were incorrectly identified by Linne.

Strombidae

No specific Strombidae were reported at urban Teotihuacan, but reputed specimens of the Panamanian Strombus (Tricornis) galeatus were reported at the Museo de Teotihuacan (Gamio 1922a), but may have been misidentified. Strombus (species unidentified) were noted at Tikal (Moholy-Nagy 1963: 69). (See Figure 32, Strombidae Distributions in Mesoamerica.)

The Panamanian S. (Strombus) gracilior was known only from West Mexico at the sites of Barra de Navidad and San Sebastian, Jalisco, and two sites in Michoacan: San Gregorio and Yurecuaro (Feldman 1968b: 170, 1974: 233, 235). S. (Lentigo) granulatus, a second Panamanian species, was also confined to West Mexico at San Gregorio and Yurecuaro, Michoacan (Feldman 1968b: 170, 1974: 235). Specimens of the Panamanian S. (Tricornis) galeatus were reported from Cuadros phase sites on the Guatemalan Pacific Coast (Coe and Flannery 1967: 78-80) and at Chiapa de Corzo (Chavez 1969: 220). From the Preclassic or Formative sites of San Jose Mogote, Tierras Largas, and Laguna Zope in Oaxaca, Pires-Ferreira (1978: 89) reported S. (T.) galeatus. Feldman (1974: 236) noted no S. (T.) galeatus specimens in coastal sites of West Mexico but did document the species at the Schroeder site in Durango and at Potrero del Calichal, Zacatecas. S. (T.) peruvianus specimens were recovered from the tombs at Las Cebollas, Nayarit, and San Sebastian, Jalisco (Furst 1966: 90-97; Feldman 1968b: 170, 1974: 235).

Five Caribbean species of Strombus were especially found in the Maya region, but some were transported to West Mexico. S. gigas, a common species, was recorded at Barton Ramie, Isla de Cancun, Uaxactun, and was probably

FIGURE 32: <u>Strombidae</u> DISTRIBUTIONS IN MESOAMERICA

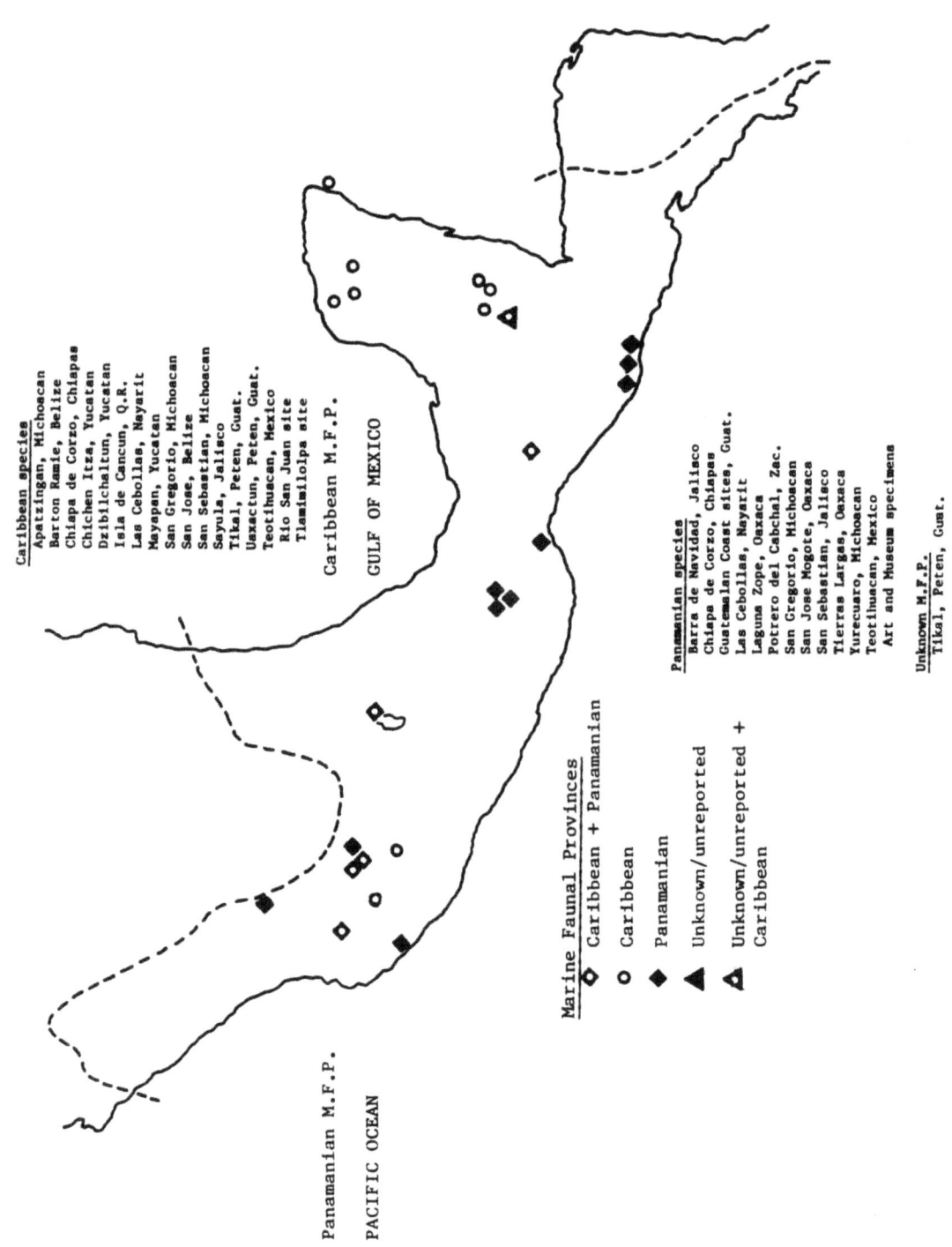

present at Tikal (Andrews 1969:37, 42, 57; Feldman 1972:125, 133; Kidder 1947; Willey et al 1965). Specimens were also noted at Chiapa de Corzo (Chavez 1969:219, 220), but apparently never reached the Valley of Oaxaca or Tehuacan Valley. In West Mexico, especially in the sites of Las Cebollas, Nayarit, and Sayula and San Sebastian, Jalisco, S. gigas imports in the form of conch shell trumpets were reported (Furst 1966:90-97; Feldman 1968b:170, 1974:232). Kelly (1947) also recorded S. gigas specimens at Apatzingan, Michoacan.

Strombus pugilis specimens from the Caribbean were found at Barton Ramie, Chichen Itza, Dzibilchaltun, Isla de Cancun, Mayapan, and San Jose (Andrews 1969:37, 52, 57; Proskouriakoff 1962:321-446; Richards and Boekelman 1937; Ricketson and Ricketson 1937; Smith et al 1950:92; Thompson 1939; Tozzer 1957; Willey et al 1965). The only examples outside of the Maya region were reported from San Gregorio, Michoacan (Feldman 1968b:170). The lesser Strombidae species S. costatus was notable only at Dzibilchaltun, Isla de Cancun, and Mayapan (Andrews 1969:37, 49, 50, 53, 57; Proskouriakoff 1962:321-446), while S. raninus was found in the Isla de Cancun midden (Andrews 1969:37, 57). S. raninus may have been among the conch shells in Maya region "structures" (Feldman 1972:125, 133). The species S. gallus is not reported from any Mesoamerican sites.

Fasciolariidae

Specimens of Fasciolaria (unspecified species), potentially from the Panamanian and/or Caribbean provinces, were found at the Ciudadela and the urban residences of Tetitla and Zacuala at Teotihuacan. The specific species F. (Pleuroploca) gigantea, which has a Caribbean habitat, was recovered from the Temple of Agriculture, Temple of Quetzalcoatl, and the Tlamimilolpa residence at Teotihuacan, while the Panamanian species F. (P.) princeps was recorded at the Xolalpan site. No other Basin of Mexico sites had Fasciolaria reported, but Porter (1956:564-565) found Fasciolaria sp. at Chupicuaro, Guanajuato. (See Figure 32, Fasciolariidae Distributions in Mesoamerica.)

F. (P.) gigantea (Caribbean) specimens were found in the Maya sites of Barton Ramie, Chichen Itza, Dzibilchaltun, Isla de Cancun, Mayapan, Tikal, and Uaxactun (Andrews 1969:38, 46, 50, 57; Kidder 1947; Moholy-Nagy 1963:67; Proskouriakoff 1962:321-446; Ricketson and Ricketson 1937; Smith et al 1961; Tozzer

FIGURE 33: <u>Fasciolariidae</u> DISTRIBUTIONS IN MESOAMERICA

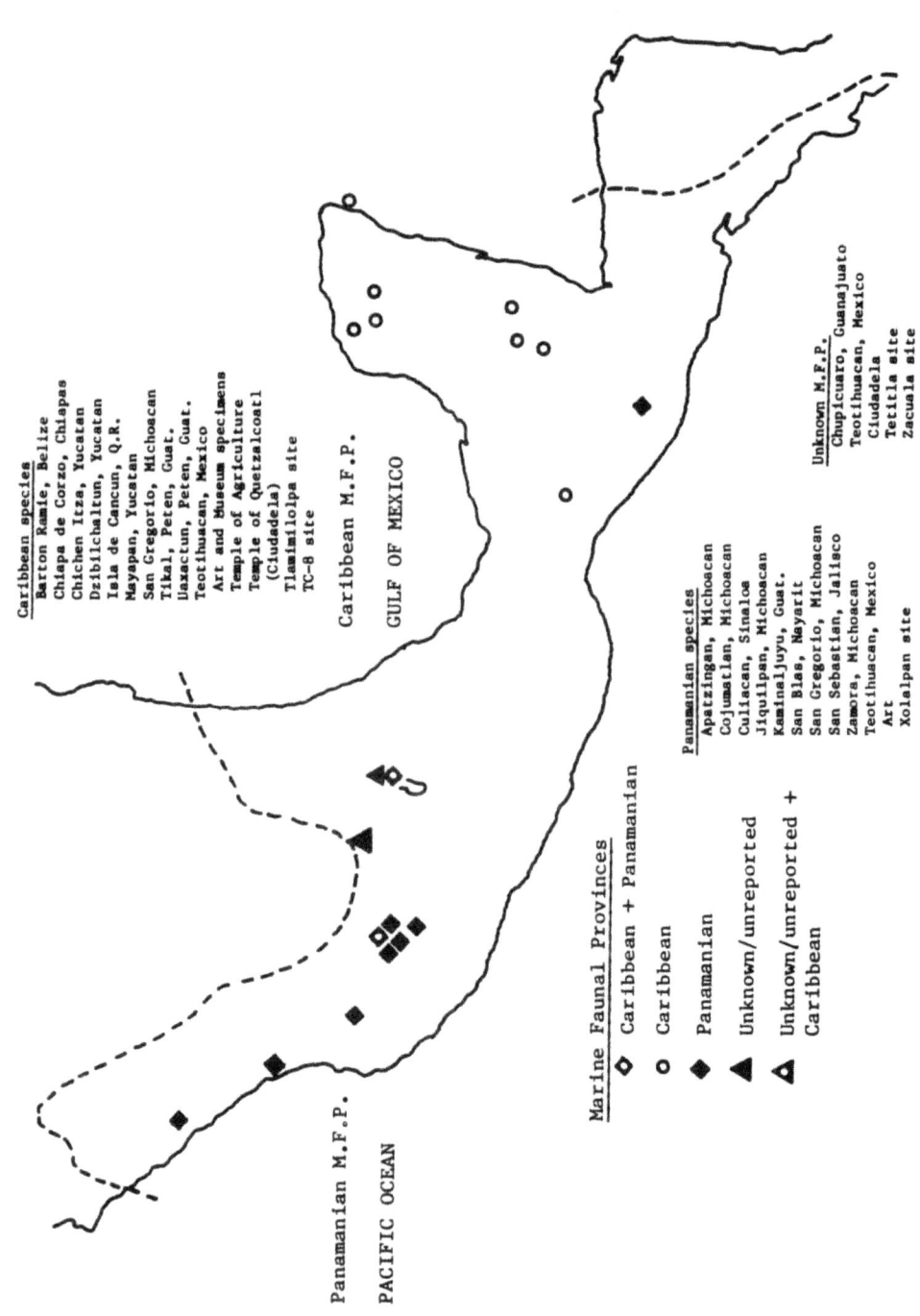

1957; Willey et al 1965). The species was also represented at Chiapa de Corzo (Chavez 1969:219), but not reported for the Valley of Oaxaca or the Tehuacan Valley. The only occurrence in West Mexico was at San Gregorio, Michoacan (Feldman 1968:169). The Caribbean species F. (P.) tulipa was noted at Dzibilchaltun, Isla de Cancun, and Mayapan (Andrews 1969:38, 46, 50, 57; Proskouriakoff 1962:321-446; Smith et al 1961). F. hunteri, also from the Caribbean, was found at Isla de Cancun and Uaxactun (Andrews 1969:38, 57, Kidder 1947).

The Panamanian species F. (P.) princeps was noted at Kaminaljuyu in the Guatemalan Highlands (Kidder et al 1946:Figure 162). Specimens were not reported for the Yucatan Lowlands, Chiapas, Valley of Oaxaca, or Tehuacan Valley. Several sites from Michoacan in West Mexico had the species represented, including Apatzingan, Cojumatlan, Jiquilpan, San Gregorio, and Zamora (Feldman 1968b: 169, 1974:234-235; Kelly 1947; Lister 1949; Long 1966; Noguera 1944). Some specimens were also found at San Sebastian, Jalisco, F. (P.) granosa specimens were recorded only at Culiacan, Sinaloa (Kelly 1945b:146). F. (P.) salmo was not reported in any archaeological contexts in Mesoamerica.

In summary, the shells of Spondylus were used in the manufacture of ornaments and jewelry, as ofrendas, and mortuary furniture. Spondylus, in the main, were not transported long distances from their marine provincial origins. Panamanian species were used predominantly in West Mexican sites, and the Caribbean species in Yucatecan sites.

The notable exception was Spondylus calcifer at Teotihuacan and the TC-8:3 site. The Strombus and Fasciolaria species, especially used as ofrendas, mortuary furniture, and as ritual shell trumpets, enjoyed wider distributions from their sources. Most notable were the Caribbean specimens in West Mexican contexts, especially S. gigas and, to some extent, S. pugilis and F. (P.) gigantea. The Panamanian species tended to be used in West Mexico and were not found in the Valley of Oaxaca, Tehuacan Valley, or Maya regions. Some Fasciolaria did reach the ceremonial center and several residences at Teotihuacan. No Strombus were specifically known at Teotihuacan, hence the shell distribution network involved the Teotihuacan state for the importation of Caribbean shells into West Mexico but not necessarily the reverse, e.g. Pacific mollusks into Gulf Coast or East Mexico.

Caribbean Strombidae were transported in other exchange networks from the Gulf of Mexico into the

American Southwest, especially to sites in southern Arizona, and to Spiro, Oklahoma (Brand 1937:301; Bell 1947:182), and sites in the American Southeast such as Etowah (Baker 1932). Gulf Coast *Strombus gigas* and *Busycon perversum* trumpets were distributed as far north as Illinois and southern Ontario (Boekelman 1936:29, 1937:295-296). The Panamanian *Strombus galeatus* was also distributed to sites in southern Arizona (Fewkes 1898:366, Boekelman 1936:27, Woodward 1936, Tower 1945). Only *Oliva* and *Olivella* had similar distributions into te Southwest from the Pacific Coast (Malouf 1940:121, Tower 1945). *Busycon perversum* and *Olivella* were reported at Tlajinga 33 in urban Teotihuacan (Storey and Widmer 1982:80-97).

CHAPTER EIGHT

MOLLUSKS AND THEIR USES

Nutrition

Along the Pacific and Gulf coasts, one of the chief uses of shellfish was as food (Safer and Gill 1982:19-21), but the shells were also used in building projects and to make utensils and ornaments (Hubbs and Roden 1964:177-178). The harvesting of bay shellfish (genera Chione, Ostrea, Pecten, and Polinices), beach shellfish (Donax and Tivela), and rocky shore shellfish (Acmaea, Haliotis, Lottia, Mytilus, and Pseudochama) resulted in the formation of extensive shell middens along both shores of the Gulf of California (Emerson 1960; Keen 1960, 1971; Feldman 1969). To a lesser extent such middens developed in sheltered bays and lagoons of the Pacific coast and along the Gulf Coast of Mexico (Thompson 1965:335).

Shellfish were and are harvested, prepared, and consumed in the local, coastal communities, but snails and freshwater genera of mussels and clams (genus Unio) were an important but adjunct food resource in central Mexico and other regions (Matteson 1959, Parmalee and Klippel 1974). Because of problems of processing and preserving the flesh of marine shellfish, this food was not normally exported to communities in the interior, although dried, salted, and pickled fish may have been transported short distances. In the Lowland Maya region, shellfish were a marginal nutritional resource during the Classic Period (Lange 1971:632, Ball and Eaton 1972:775). The heaviest shellfish exploitation apparently occurred during the Archaic and Preclassic to judge from midden cultural associations. Likewise, in the Pacific Coast Ocos area, Coe and Flannery suggest that the local population did not heavily rely on molluskan foods (1967:80) but further north, shellfish middens abound (Lorenzo 1955:47), especially in mangrove swamps.

Utilitarian and Construction

In addition to their obvious nutritional value, the shells of the marine Mollusca were important raw materials for implements and construction. Large Pelecypoda (clam) shells made useful containers and receptacles, and, hafted or unhafted, made simple hoes and digging tools (Leechman 1949; Hubbs and Roden 1964:179-180; Safer and Gill 1982:30-34). In Prehispanic times along the Pacific coast of Guatemala and the Gulf Coast of Mexico and the Yucatan Peninsula, common Mollusca shells were used in construction fill and more importantly, burned for lime to make cement or mortar and plaster (Hubbs and Roden 1964:180; Pollock 1965:396, 404; Shook 1965:186; Keen 1971:96; Safer and Gill 1982:29-30).

Shellfish Dyes

Shellfish of the Gastropoda genera Murex, Purpura, and Thais served as sources of dye in communities on the Pacific Coast of Mesoamerica and Central America (Martens 1899, Nuttall 1909, Jackson 1917:23-29, Gerhard 1964; Safer and Gill 1982:27-29). The Panamanian Marine Faunal Province has two species of Murex, one of which has three subspecies (Keen 1971:514-516). The most common is Murex (Murex) elenensis Dall, 1909, whose habitat extends from Baja California to Ecuador. Murex (M.) recurvirostris recurvirostris Broderip, 1833, is found from southern Mexico to Ecuador. M. (M.) recurvirostris lividus Carpenter, 1857, is found in the Gulf of California south to Mazatlan, and M. (M.) r. tricoronis Berry 1960, is confined to the Gulf of California.

The genus Purpura has two species, P. columellaris Lamarck, 1822, found from the Gulf of California to Chile, and) P. pansa, Gould, 1853, found from the Gulf of California to Colombia (Keen 1971:552-553). Thais has five subgenera and a total of nine species (Keen 1971:548-550). Five species have habitats from Baja California to Peru or Chile: Thais (M.) speciosa Valenciennes, 1832; T. (M.) triangularis Blainville, 1832, T. (S.) biserialis Blainville, 1832, T. (T.) kiosquiformis Duclos, 1832; and T. (T.) planospira Lamarck, 1822. Thais (Vasula) melones Duclos, 1832, ranges from the Gulf of Tehuantepec to Peru.

All species within these three genera secreted a milky fluid which, on exposure to air turns to purple or purplish-red. Ethnographically, the most important because of size, quantity of fluid, and ease in gathering was Purpura pansa (Gerhard 1964:178-179, Hubbs and Roden 1964:180). Important loci of collection included the West Mexican coast from Ostula to Pomaro in Michoacan, at Acapulco, Guerrero, and in southern Oaxaca from Puerto Angel to Tehuantepec (1964:179, 182-185, Map 1).

The three shellfish dye genera are also represented in the Caribbean Marine Faunal Province but were apparently little exploited in Mesoamerica (Gerhard 1964:178). Murex is represented by four subgenera and a total of six species (Warmke and Abbott 1962:104-106). The majority are found from Florida through the West Indies, but two are found in the Gulf of Campeche and the coasts of the Yucatan Peninsula: Murex (Phyllonotus) pomum Gmelin, 1791, the "Apple Murex," and M. (Chicoreus) brevifrons Lamarck, 1822, the "West Indian Murex." The Purpura patula Linnaeus, 1758, is the only Purpura found in the Caribbean and is associated with the Gulf of Campeche and Yucatecan Coasts (Warmke and Abbott 1962:107). Three species of Thais inhabited the same zones (1962:107-108): Thais (Stramonita) rustica Lamarck, 1822; T. (S.) haemastoma floridana Conrad, 1837; and T. (Mancinella) deltoidea Lamarck, 1822. Of these the Murex pomum, M. brevifons, and Purpura patula were potential sources of purple dye. The Nuttal Codex has a purple paint which may have been derived from Purpura (Nuttall 1909:380-381, Jackson 1917:25).

Shell Artifacts

Gastropoda and Pelecypoda shells provided the raw materials for the manufacture of fishhooks (Stresser-Pean 1971:590). Most importantly, shell was cut and ground into ornaments and jewelry including bracelets, discoidal beads, earplugs, gorgets, necklaces, nose ornaments, pendants, plaques, rings, "tinklers," and tubular beads used for personal adornment and in ritual and funerary activities (Starr 1897:57; Woodward 1936:117-118; Lowe and Mason 1965:215; Rands 1965:577; Safer and Gill 1982:74). Such manufacture occurred throughout the Yucatan Lowlands, Pacific Coast, and, to a lesser extent in the Guatemalan Highlands (Caso 1965:900, 902, 906, 909, 911; M. Coe 1965:697, 700; Roys 1965:671-672; Woodbury 1965:177). Shell ornaments were also made and used throughout northern Mesoamerica, especially in West Mexico (Bell 1971:705-706, 729; Chadwick 1971:665, 676; Harvey 1971:631; Kelley 1971:770-771, 780-782, 790; Meighan 1971:764; Stresser-Pean 1971:591), but also in the Meseta Central (Cook de Leonard 1971b:216, Piña Chan 1971:175) and the Sierra (MacNeish 1971:575). Artifacts and ritual materials were also derived from Caribbean shark teeth and stingray spines (Borhegyi 1961:273).

More elaborate artifacts were fashioned from shell by Mesoamerican artisans. Shell served as a backing for some jade earplugs (Rands 1965:565) and was used with jade or jadeite and obsidian to make mosaic plaques and ceremonial knife handles (Caso 1965:899, 905; Noguera 1971:262). Small natural shells such as the Gastropoda genera Astraea and Oliva were drilled and strung as decorations and "tinklers" on shirt and kilt fringes (Mahler 1965:584; Weitlaner Johnson 1971:311-312; Storey and Widmer 1982:33, 87). Incised scenes decorated shells found at Matacapan and Piedras Negras, among other sites (M. Coe 1965:711; W. Coe 1965b:466; Stresser-Pean 1971:594-595). Among the Postclassic Lowland Maya both cacao "beans" and red Spondylus shells and beads served as mediums of exchange (Borhegyi 1965a:6; Roys 1943, 1965:670). Both modified and unmodified marine shells were used in ceremonial caches and ofrendas and in tombs as grave furniture especially in the Peten of Yucatan and West Mexican shaft tombs (W. Coe 1965b:465, 468; Willey and Bullard 1965:377; Furst 1966:90-97; Noguera 1971:260). Perhaps one of the most interesting uses of conch shell was reported by Siliceo Pauer (1925:210-211), who noted that artificial human teeth were fashioned by cutting and grinding. This may be early evidence of restorative dentistry in Mesoamerica.

Shell Trumpets

World wide, Strombus shells have been used to make "shell trumpets," and such artifacts ethnographically are important in social ceremonies and religious rituals (Safer and Gill 1982: 138-152, 168, 174-175). Larger Strombus shells from both the Panamanian and Caribbean Provinces were modified into conch shell trumpets in Prehispanic times and used as funeral offerings in the Maya Lowlands (W. Coe 1965a:601) and Guatemalan Highlands (Kidder, Jennings, and Shook 1946:147; Hubbs and Roden 1964:180; Woodbury 1965:177). Strombus trumpets made from species from both marine provinces were found in tombs in the West Mexican states of Michoacan (Chadwick 1971:665), Colima, Jalisco, and Nayarit (Furst 1966:90-97; Bell 1971:717, 720, 726, 728, 734, 748-749; Kelley 1971:770-771). Jackson (1917:47-50, 55-56) early contended that these trumpets were made from West Mexican and Peruvian genera.

The shaft tombs in the Early Classic site of Las Cebollas in southern Nayarit bear special mention. Furst (1966:94-97) recovered 125 complete or fragmentary conch shells of which 120 were Turbinella angulatus Solander (Xancus angulatus S.), the "West Indian Chank" native to the Caribbean (Morris 1951:209-210), and four were Strombus gigas Linnaeus 1758, the "Queen Conch," also native to the Caribbean (Morris 1951:165-166, Warmke and Abbott 1962:88). One Panamanian conch, Strombus peruvianus (Strombus (Tricornis) peruvianus) Swanson, 1823, whose habitat is from coastal Nayarit south to northern Peru, was also associated (Keen 1971:421).

Of the 125 specimens, 111 were "end blown" trumpets, an additional seven fragmentary specimens were possible trumpets, and seven conchs were unmodified. Eighty-five had from one to four drilled holes suggesting the use of carrying straps or the suspension of ceremonial objects such as small shells or feathers. Seven Turbinella had incised and carved decoration, while one additional Turbinella bore "... green and yellow fresco paint in combination with an incised pattern which resembles the most common Teotihuacan shell decoration ..." (Furst 1966:95, Pl. 45). All five Strombus were undecorated trumpets.

One Turbinella contained forty-six tubular shell beads, and another trumpet had two "ornaments of unknown significance" inside (1966:96-97). Twenty-six complete and forty-eight fragmentary shell bracelets and two shell rings were also found in the shaft tombs. A radiocarbon date of ca 100 A.D. was obtained from one Turbinella specimen.

Long's excavation of a shaft tomb at San Sebastian in northern Jalisco produced Strombus gigas specimens radiocarbon dated to 140 B.C. (Bell 1971:726). The tomb also contained a Murex nigritus Philippi, a Panamanian Gastropoda dated to 400 A.D. This latter specimen was more likely Murex (Murex) recurvirostris Broderip, 1833, which has three recognized subspecies inhabiting the West Mexican coast (Keen 1971:514-516). From Lake Chapala, Jalisco, a Fasciolaria princeps Sowerby (Fasciolaria (Pleuroploca) princeps Sowerby, 1825) shell trumpet was reported by Bell (1971:728). She attributed the specimen to the Caribbean, but it is a Panamanian gastropod (Keen 1971:611). The shell may have been misidentified, as there is a Caribbean twin species F. (P.) gigantea Kiener, 1840 (? = F. papillosa Sowerby, 1825).

The incidence of Caribbean specimens in West Mexican contexts since the Late Preclassic and especially the Early Classic suggests a long-distance exchange network probably involving ideology as well as iconography (Furst 1966:153-170, Bell 1971:748-749). Such a Gulf Coast/Caribbean-West Mexican connection involved the Classic Teotihuacan polity. Both the distance and quantities of shell may not be surprising. For example, the "Queen Conch," Strombus gigas is presently a commercial shell exported from the Bahamas for cutting into cameos, while the scrap material is ground into powder for manufacturing porcelain. Scrapers, chisels, and other tools are made from the shell, and the flesh is eaten (Morris 1951:166). The export of this conch from the Cacaios Islands from 1904-1956 exceeded 87.8 million specimens (Doran 1958:389).

Safer and Gill, who wrote Spirals from the Sea: An Anthropological Look at Shells (1982), have also commented on the importance of mollusks as food, construction material, a source of dyestuffs, and a raw material resource for artifacts such as jewelry. In addition, they noted the significance of marine shells cross culturally as means of calculating wealth and as status symbols, especially of social and/or political rank and associations with rite de passage events (1982:49-72, 85-108). Wealth, they observe, can be calculated in terms of unmodified shells or shells which have a hole drilled in them for hanging, notably cowries (Cypraea) and "tusk" shells (Dentalium). The shell necklaces which figure so prominently in the Kula Ring ceremonial exchange system of the Trobriand Islanders of Melanesia were made of Spondylus (sp.) shell disks (1982:69-70).

In summary, marine mollusks served many uses in Prehispanic Mesoamerica, especially in the production of jewelry and ornaments, and as ceremonial and ritual offerings. Shells provide important evidence of cultural contact because mollusks live in restricted habitats (Jackson 1917:23-26, 47-60). As a source for radiocarbon dates, shell from freshwater and marine habitats are useful (Evans 1969:171-173), but, as in dendrochronology, chronologically "old" materials may be used in "younger," more recent contexts. Shell artifacts, like some ceramics ("Thin Orange" Ware and Monte Alban II-IIIA urns, for example) and other material culture, can acquire heirloom status and be associated in more recent archaeological contexts (Kolb 1984a:209, 1986:160). This would account for the 140 B.C. Strombus gigas trumpet in association with a 400 A.D. Murex nigritus in the San Sebastian, Jalisco tomb (Bell 1971:726).

CHAPTER NINE

THE SHELL PROCUREMENT NETWORK

Introduction

The thorny oyster, Spondylus, and the conchs Strombus and Fasciolaria occupy slightly different marine ecological niches, but species within each genera could be, and were, obtained along the Gulf Coast and Campeche Bank of the Caribbean Marine Faunal Province, and from the Gulf of California south to the Isthmus of Tehuantepec in the Panamanian Marine Faunal Province. In Prehispanic times, Spondylidae apparently served primarily as raw material for the manufacture of ornaments and jewelry, whereas the conchs were especially modified into shell trumpets for ritual and ceremonial uses, as well as for inclusion in mortuary contexts. Spondylus and Strombus were especially important as elite insignia in ceremonial centers and in burials in sites in the sierra and coast of the Central Andes since 1100 B.C. (Paulsen 1974). In Mesoamerica these genera were significant in the Yucatan Lowlands, but also in the Guatemalan Highlands, West Mexico, and the Meseta Central since at least 800 B.C. (Davidson 1980). During the Neolithic of the Balkans and Central Europe (ca 4000 B.C.), Aegean Spondylus (S. gaederopodus Linnaeus, 1758) served as raw material for the production of high status "prestige goods" (Shackleton and Renfrew 1970).

In Mesoamerica, the system or procurement network, involved obtaining, manufacturing or modifying Spondylus and the conchs into "prestige goods," and subsequently distributing them into regions often far distant from the coasts. In the case of Strombus trumpets, the system required the transport of shells from the Caribbean province to sites in West Mexico (Furst 1966:153-170; Bell 1971:720, 726). The following discussion will focus on the Classic Teotihuacan Period, ca 50-750 A.D., but with emphasis on the waxing and waning of the "empire," ca 300-650 A.D. The minimal coast-to-coast distances from the Caribbean sources to Jalisco and other West Mexican sites was 550 kilometers, while Teotihuacan and the Gulf Coast were minimally 235 kilometers apart, and the distance from Teotihuacan to Acapulco on the Pacific Coast was at least 325 kilometers. Drennan (1984a:28, 1984b:105) believes that 275 kilometers was the absolute maximum distance for the "profitable" overland transport of food staples in Mesoamerica. The import distance for Gulf Coast shells to Teotihuacan was approximately that of Drennan's food import distance, but the Teotihuacan to

Pacific Coast distance would have far exceeded the food import distance.

Some Exchange System Parameters

By institutionalizing segments of long-distance commerce, custom replaced chance, and it was possible for the state and its affiliated, licensed "imperial" agents to maintain ties over a considerable distance without the risk of competitive intrusion (Vance 1971:21). A vast array of luxury imports came into the Supra-Regional center and the Teotihuacan state from frontier regions and "foreign" areas. Among the imports from tropical areas was copal (copalli), a resinous incense derived from trees of the Family Bruseraceae (Elophrium jorultense and E. aleoxylon), used in incensarios and candeleros during religious rituals and perhaps social ceremonies as well (Cook de Leonard 1971b:221). Bark paper (amatl) derived from trees of the genera Ficus and Ehretia tinifolia also was imported (Cook de Leonard 1971b:221). Amber and tropical bird feathers including quetzal came from the regions of Chiapas and Guatemala (Barlow 1949:98, Miles 1965:279-280). Other luxury goods included cacao from the Gulf and Pacific Coasts (Barlow 1949:8, 32; Miles 1965:279; Sanders and Price 1968:168-169; L. Parsons and Price 1970:20-21; Silva-Galdames 1971:42, 52; Litvak King 1978:120).

Cotton, grown in the tierra templada of the Tarascan frontier and Morelos region (Barlow 1949:8, 32; Cook de Leonard 1971b:221-222, Parsons et al 1982:381), was a significant import. Cotton produced in the Guatemalan Pacific slope region might also have been an import (Miles 1965:279), but less likely because of the distance involved. Besides copal, cacao, paper, and cotton, Sanders and Webster (1978:287) believed that tropical fruit, root crops and honey came from frost-free regions of Morelos. Millon (1981:227-228) stipulated that "blue-green mineral stones," "white jade," flint, cinnabar, and hematite were obtained from the Chalchihuites region of Zacatecas, ca 250-500 A.D. New Mexican turquoise, possibly from Cerrillos, may have been indirectly obtained through the Chalchihuites region. Alpha "Thin Orange" ceramics produced in southern Puebla or nearby regions constituted another major import (Kolb 1973c:337-340, 1977:535, 1982:195-197, 1984a:214-217). Marine shells from both the Gulf and Pacific Coasts were also imported into the Teotihuacan state (Kolb 1973a, Millon 1981:227).

While "exotic" and luxury goods and raw materials were the major imports, the Teotihuacan state exported obsidian artifacts produced in "state controlled" workshops (Charlton 1978, 1984; Sanders and Webster 1978:287; Spence 1984; Kolb 1986:172-175). Ceramics produced at Teotihuacan, especially Cylindrical Tripod Support Vases and Copoid Wares, were among the tangible exports (Kolb 1987a, 1987b). Religious iconography and ideology were also Teotihuacan "exports" (Pasztory 1978).

The Teotihuacan polity had the attributes of a state-level society (Sanders and Price 1968:140-141, 207-208; Sanders and Webster 1978:274-275; Sanders et al 1979:400-402; Millon 1981:228-235). The state became an "empire" primarily because of its commercial rather than military capabilities ca 300 A.D. The "evolution" of multi-national corporations in our present economic climate offers a parallel. The state controlled large-scale movements of raw materials, foodstuffs, and finished products, and the Great Compound served as the major marketplace of the Supra-Regional Center. Because of the quantities of imports, a professional merchant class undoubtedly existed. However, there is no direct evidence to suggest "ports of trade," such as those known in late Postclassic Aztec/Culhua-Mexica and Lowland Maya cultures. In the next sections I shall consider the merchants, potential "ports of trade," and commercial routes.

Professional Merchants

A number of authors writing about the trade and mercantile systems of the Preclassic Olmec, Classic Teotihuacanos, and Postclassic Toltecs have frequently used the Aztec/Culhua Mexica pochteca as a model for elaborating exchange and distribution systems in pre-Aztec Mesoamerica (M. Coe 1965; Flannery 1968; Grove 1968b, 1968c; Sanders and Price 1968:122; Sanders et al 1979:400-402; Kolb 1977:535, 1984a:218, 1986:186-187; Mountjoy 1978:135). Carrasco (1971:350) has summarized the significance of merchant-traders in Prehispanic Mesoamerica:

> Large-scale trade of the type that could create great wealth was government controlled. Traders used their accumulated surplus for feasts and distributions that enhanced their social status, but checked further accumulation. Their wealth was also subject to confiscation by the tlatoani [noble rulers]. Large-scale, long-distance traders were important economic and political agents of the rulers, and some traders had themselves the status of noblemen; but in no case was

there created a threat to the ruling class based on the political control of land and tribute.

Therefore, the importation of "exotic" and luxury goods was under the control of a specialized merchant group affiliated with the state. Such merchant groups were operating during the Preclassic as well as the Classic Periods (Bittman and Sullivan 1978:221, Millon 1981:210, 219, 226). It is not known if the professional merchants of the Classic Period had socioeconomic attributes identical to those of the late Postclassic Aztec/Culhua Mexica, but the basic characteristics were probably similar.

The term pochteca was a generic name for "merchant," some of whom were associated with the state as "imperial" merchants, while others were independent "private sector" merchants. During the Postclassic, especially from 1428-1520 A.D., professional merchants were the purveyors of luxury goods and "exotic" materials for the nobility and elite of Aztec/Culhua Mexica society. Some merchants were pochteca affiliated with the state and "empire" while other merchants were regionally-based entrepreneurs (Acosta Saignes 1945, Barlow 1949, Berdan 1982:35, Bittman and Sullivan 1978, Feldman 1978:138-139, Mountjoy 1978:134-137, Sahagun 1959:32). The pochteca were organized into guilds (or sindicatos), were concentrated in specific calpulli within the cities and large towns, and occupied a social position, along with luxury artisans, immediately below that of the nobility. Often, as agents of the state, they engaged in long-distance trade, and paid tribute in goods to the rulers of the class of nobles (tlatoani). As quasi-nobles (e.g. "high status" individuals) the pochteca enjoyed special privileges, including having their own temples and deities, and law courts. At least some pochteca groups were not hereditary. The guilds quickly dissolved after the Spanish conquest, but long-distance and local trade continued (Berdan 1982:46, 176).

Our knowledge of the pochteca comes primarily from post-1520 A.D. codices and chronicles (Kolb 1986:187). The major source was Sahagun's Florentine Codex, Book 9: The Merchants (1959), written before 1569, which has most frequently provided the model for explaining the origin, organization, and functions of the pochteca and other merchants. The particular kinds of pochteca were: 1) oztomeca or "walking merchants" or "those who went out on the road," 2) tecunenenque or yaque, the "travelling lords" who were royal of "imperial" trade officials, 3) nahualoztomeca, the "merchant-spies and warriors," 4) teyahualoani, "besiegers," and 5) yaoc calaquini, "enterers into battle." Therefore, some merchants also served as spies and conquerors as well as "imperial" agents of the state (Sahagun 1959: 7-8). "Imperial" merchants also conducted their own "private sector" business and purchased and sold other goods (1959: 8, 18, 21-22), but slave-trading was reserved for merchant "chiefs" called teyacanque or pochtecatlatoque (1959: 18, Chapman 1957: 121). Only "imperial" merchants acting as agents of the monarch were permitted to travel to exclusive, reserved "trading areas" (Sahagun 1959: 17-18). These professional merchants knew the trade routes, the regions, the markets, local languages, and the needs of the people "as no one else did" (Bittman and Sullivan 1978: 213).

Merchant caravans were led by pochtecatlatoque and included various pochteca and non-pochteca, slaves and burden carriers/porters (tlamemes). The oztomeca and nahualoztomeca acted in commercial transactions and obtained detailed intelligence for the "empire." The state became increasingly involved in these economic transactions which led to more state regulations and to a near-monopoly (Bittman and Sullivan 1978: 217). During the Classic Teotihuacan Period, we may assume that at least the first two categories of pochteca, the oxtomeca ("private sector" merchants) and tecunenque ("imperial" merchants), were the kinds of commercial specialists affiliated with the Teotihuacan state and "empire." The latter as "imperial" merchants would have had access to exclusive trading areas or "ports of trade" or "free trade" zones. Again, a Postclassic model may be applicable in the Classic setting.

"Ports of Trade"

"Ports of trade" were relatively neutral locations, easily reached by merchant caravans of tecuneneque and their burden-carriers/porters (tlamemes) or by canoe transport, which served as meeting places of non-local "foreign" merchants who conducted economic transactions in luxury goods and other items (Chapman 1957: 115, 128, 1959: 3; Berdan 1978: 187). A coastal or riverine locus was not necessarily implied, but such loci would have excellent transportation access and appropriate storage facilities. During the Aztec Postclassic, the region and town of Anahuac-Xicalango and associated communities on the Veracruz Gulf Coast (Chapman 1957: 135-141, Berdan 1978: 191) and the province of Anahuac-Ayotlan/Xoconusco on the Chiapan-Guatemalan Pacific Coast (1957: 141-142) were "free trade" loci for the exchange of luxury goods, ceramics, "exotic" foodstuffs, and unique raw materials. The town of Tochtepec (Tuxtepec) in present-day northeastern Oaxaca, was a "frontier" trading center established by merchants from Tlatelolco a Tenochtitlan (Culhua Mexica) ally in the Basin of Mexico (Chapman 1957: 123, 137).

A number of Postclassic Maya "free trade" centers served the same basic function. Of special note were Cozumel, an island off the east coast of the Yucatan Peninsula, a second center in the Bay of Chetumal in present-day Belize, and a third cluster in the Gulf of Honduras at the confluences of the Rios Sarstoon, Dulce, Montagua, and Ulua (Chapman 1957:145-146; Cardos 1959; Tortellot and Sabloff 1972; Sabloff and Freidel 1975; Sabloff and Rathje 1975a, 1975b; L. Parsons 1978:28). In the interior of the Yucatan Lowlands, Acalan served a similar function as Tochtepec (Chapman 1957:142-145), and Cachi in northeastern Yucatan was a late Postclassic Maya "port of trade" (Roys 1965:671). No other Postclassic "free trade" centers have been identified on the Pacific Coast except Anahuac-Ayotlan/Xoconusco, but such a center could have existed in choice areas such as the Bahias de Banderas, Petacalco, and/or Acapulco, among others. Archival and cartographic sources and archaeological data do not support the existence of an Acapulco center during the Postclassic or during earlier times, nor is there evidence for "free ports" at the mouth of the Rio Balsas/Bahia de Petachalco or at Puerto Vallarta/Bahia de Banderas on the Mexican Pacific Coast.

Of the known Prehispanic "free trade" centers, only Cozumel has been investigated and studied in detail (Sabloff and Freidel 1975; Sabloff and Rathje 1975a, 1975b). This center was a "pivotal node" in the Caribbean Precolumbian Maya long-distance commercial network (Cardos 1959). Sabloff and his colleagues have identified warehousing, packaging and transportation facilities, administrative offices, religious areas, domestic residences, and craft workshops at Cozumel. Undoubtedly Xoconusco and Xicalanco, and, to some extent, Tochtepec, had similar features and facilities. Armytage (1953) and Thomas (1956) have delineated parallel attributes in historic Caribbean "free trade" loci.

The items obtained at Anahuac-Xicalanco included various "green stones," jade, jadeite, turquoise mosaics, sea shells, tortoise shells, bird feathers, and the skins of wild animals (Sahagun 1959: 17-19). Among the shells obtained were "...large red sea shells, and red coral shells -- the very red (ones), and flower-colored shells -- the very yellow (ones), and tortoise shell cups ..." (1959:19). The red shells were identified as tapachtli in the Codex Mendoza (Clark 1938[2]:116), and were undoubtedly the Spondylus americanus Gmelin, 1781, the Caribbean species.

In summary, I propose that the Teotihuacan polity had a group of professional merchants similar to the oztomeca and tecuneneque. Although no direct evidence exists for Teotihuacan "ports of trade," Teotihuacan had commercial ties to Kaminaljuyu in the Guatemalan Highlands, to Matacapan on the Gulf Coast, and to other locations on the Pacific Coast, possibly in the vicinity of Acapulco Bay and the mouth of the Rio Balsas (Sanders 1978:37, M. Coe 1965:683, Litvak King 1978:120).

Potential Teotihuacan Commercial Routes to the Coasts

Significant routes connecting the Gulf Coast and Meseta Central and Valley of Oaxaca were developed by the Olmec for the importation of "exotic" materials and the exportation of finished products, iconography and ideology (Flannery 1968; Grove 1968a, 1968c; Pasztory 1978:3). At least some of these routes were used by Teotihuacan merchants during the Classic Period. The "Eastern Basin of Mexico Route" (Route One) and "Western Basin of Puebla-Tlaxcala Route" (Route Two) were involved in the importation of goods from the southern regions of th Meseta Central (Morelos and Puebla) into the Teotihuacan polity (Kolb 1984a:217-219, 1986:175-181). These routes would also have been used in the export of Teotihuacan obsidian to the south and southeast. However, there is no evidence to suggest that quantities of marine shell were either imported or exported along these avenues. (See Figure 6, page 8.)

Gulf Coast Routes A-G

There are seven potential commercial routes from the Teotihuacan Supra-Regional Center to the Gulf Coast. I shall designate these as routes A through G. (See Figure 34.) "Gulf Coast Route A" proceeded north from the Teotihuacan Valley and Basin of Mexico overland toward the obsidian sources near Pachuca then into the drainage basin of the Rio Tempoal and Rio Panuco to La Huasteca and the Laguna de Pueblo Viejo and Laguna de Tomiahua. A second potential passage, "Gulf Coast Route B," came overland from the Basin of Mexico toward present-day Tulancingo then via the drainages of the Rio Calzones to the coast north of Classic El Tajin. An alternate "Route C" left Teotihuacan via the East or Eastern Avenue, passed through the upper Teotihuacan Valley near Otumba and continued eastward across the low range of hills separating the Teotihuacan Valley and Plains of Apam (Apan) in the northern Basin of Tlaxcala, possibly to the Teotihuacan-controlled obsidian production center at Tepeapulco. "Route C" passed through the mountains to

FIGURE 34: GULF COAST TO HIGHLANDS COMMERCIAL ROUTES DURING THE CLASSIC TEOTIHUACAN PERIOD

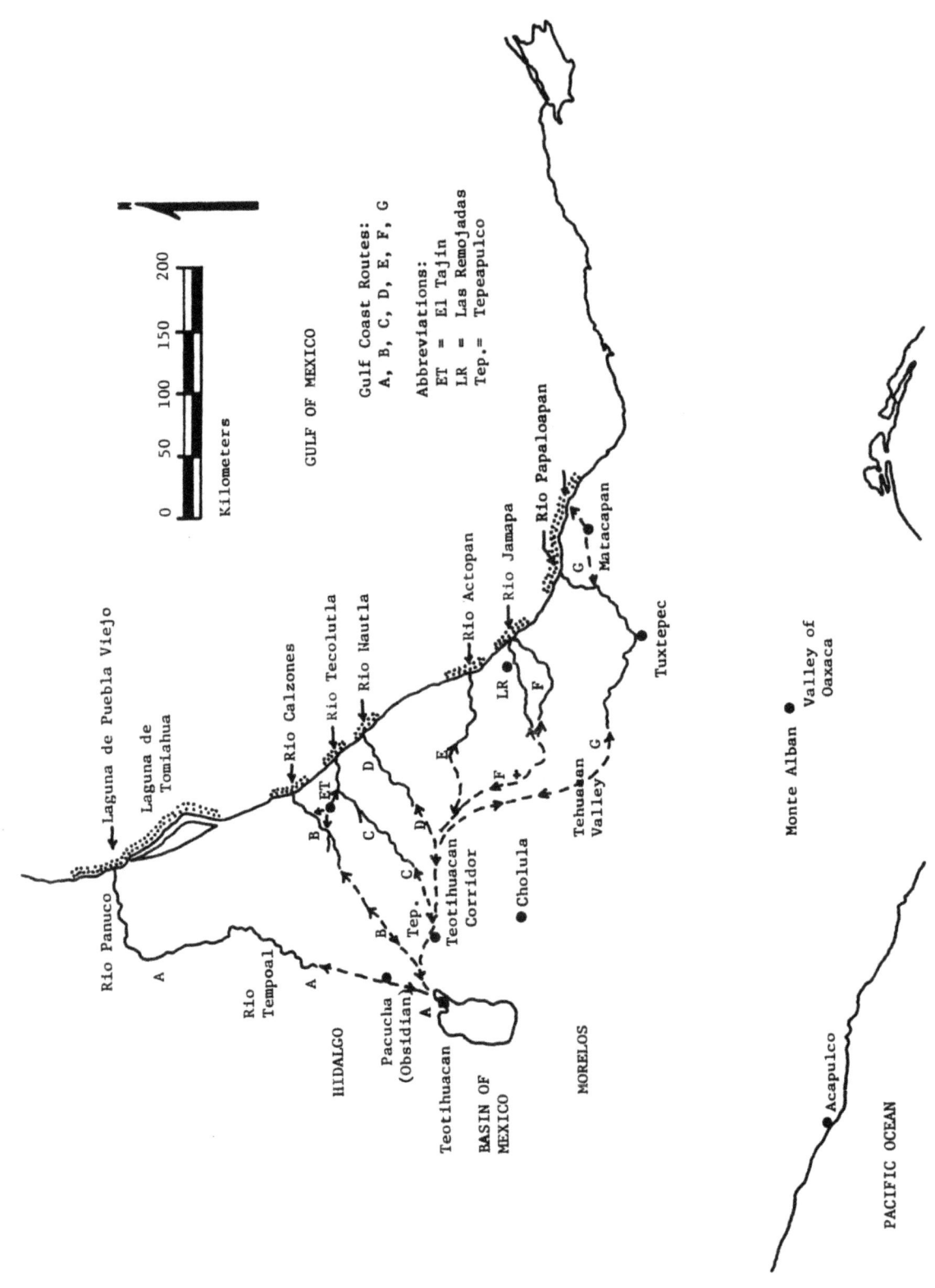

the east and entered the Rio Tecolutla drainage basin hence to the coast south of Classic El Tajin.

The initial segment of "Route C" was the western portion of the "Teotihuacan Corridor," with Tepeapulco located at the northern edge of this passage and Calpulalpan situated at the southern in the Tenanyecac Region. this "corridor" was in use since at least ca 200 B.C., and at least one Classic Period El Tajin settlement was founded by migrants from the Papantla Region (Garcia Cook and Trejo 1977:59, Kolb 1986:179-181). Over twenty Teotihuacan-affiliated settlements have been identified in the northern Basin of Tlaxcala (Garcia Cook and Trejo 1977:59; Charlton 1978, 1984). The corridor continued east through the northern Basin, then turned southeast, passing east of Cerro La Malinche.

"Gulf Coast Route D" followed the "Teotihuacan Corridor" to the area northeast of La Malinche where the passage continued overland to the northeast into the drainage basin of the Rio Nautla hence to the coast. Alternate "Route E" also used the corridor to the east of La Malinche then continued northeast through the Tepeyahualco region via modern Perote, Jalapa, and Coatepec into the Rio Actopan drainage to the Veracruz coast north of the modern city of the same name. Portions of Routes D and E were used by Cortes in 1519 during his march to the Basin of Mexico.

"Gulf Coast Route F" followed the corridor, passed southeast of La Malinche and between Volcan Citaltepetl and Pico de Orizaba into the basin of the Rio Cotaxtla and/or Rio Jamapa to the Veracruz coast. This route possibly involved the site of Las Remojadas located north of the Rio Jamapa. "Route G" likewise used the corridor to the east of La Malinche, but, instead of turning east, continued south through the Tehuacan Valley to the vicinity of modern Tuxtepec (the Aztec/Culhua Mexica Tochtepec), then northeast into the Tuxtla Region of the Gulf Coast.

The roles of the Classic Period site of Cerro de las Mesas near the Rio Blanco and the Late Preclassic site of Tres Zapotes on the Rio Papaloapan are not known. However, the Classic site of Matacapan, located on the coastal plain between Lake Catemaco and the slopes of Volcan San Martin Tuxtla, was probably a focus of Teotihuacan commerce (M. Coe 1965:683, 704; Drennan 1984a:36). Matacapan showed Teotihuacan "influence" in its axial orientation slightly east of north, in its architectural style, and in ceramics, which included <u>Alpha</u> "Thin Orange" (Kolb 1986:195). Potentially, Matacapan was a commercial center or "port of trade" perhaps similar in function to Kaminaljuyu in the Guatemalan Highlands (Sanders 1978), but it may have been a Teotihuacan Provincial Center on the Gulf Coast. It is located in the region of Anahuac-Xicalango and associated late Postclassic communities which were a "port of trade" zone for the Aztec/Culhua Mexica (Chapman 1957: 135-141, Berdan 1978: 191).

The Matacapan site, because of its Teotihuacan "connections," was selected for extensive and limited intensive archaeological examination by Robert S. Santley and his colleagues (Santley et al 1984, 1985; Santley 1984: personal communication). Because their data is still being analyzed, the following statements should be regarded as preliminary assessments of three field seasons, 1982, 1983 and 1984. The researches of earlier investigators and Santley's own initial archaeological reconnaissance and surface collecting suggested that there was a localized Teotihuacan barrio, Area B, located directly west of the Main Plaza, at the Matacapan site (1984: 22). The extent of Classic Teotihuacan (or Teotihuacanoid) occupation was between 2.0 and 5.0 km^2 and included a number of mounds with Teotihuacan architectural elements, such as the talud and tablero. The population was estimated at a minimum of 3,600 to a maximum of 7,200 (1984: 23). Mounds 1 and 2 were temple platforms in the Teotihuacan style, while two extensive mounds, Numbers 2 and 22, supported multi-roomed residences (1985: 1-2). The area of Teotihuacan occupation contrasted markedly with the Kaminaljuyu site in Guatemala, where Teotihuacan occupation was confined to the Mound A/Mound B complex.

Santley's working hypothesis was that the Teotihuacan barrio "was occupied by an enclave of merchants from the Basin of Mexico metropolis [urban Teotihuacan] and had control over the long-distance exchange of obsidian and other exotics such as cacao, semi-precious stone and tropical feathers which was a primary impetus behind the contact process" (1984: 77). His data suggested to him that there was a "different process of contact" in comparison to Kaminaljuyu. "Thus, rather than a small barrio of foreigners that presumably had a commercial function, Matacapan may have been physically conquered by Teotihuacan" (1984: 78) [emphasis mine].

Obsidian, clay figurine, and ceramic workshops were reported (1984: 19, 77-79; 1985: 3-4, 28-31, 51-52). Much of the obsidian in the form of blanks and blades came from the Pachuca source in the Meseta Central, controlled by Teotihuacan, while only some obsidian came from El Tajin, Veracruz, an affiliated Teotihuacan polity (1984: 78). Santley believes that the Teotihuacan barrio was built as a single construction event and occupied by

Teotihuacan peoples during the Middle Classic (the Early and Late Tlamimilolpa phases at Teotihuacan, ca 300-450 A.D.) but that the "heaviest occupation was during the Late Classic" (Early and Late Xolalpan phases and early Metepec phase, ca 550-700 A.D.) (1985:13-14). Through time, the barrio occupants probably adopted local customs, material culture, and foods (1985:5). During the 1984 field season, a residence and one of ten ceramic workshops were excavated (Santley 1984:personal communication). Santley earlier contended (1984:19) that "most of the Cylindrical Tripod Support Vessels at Matacapan are imitations of Teotihuacan vessels rendered in local paste, often in Tuxtlas Fine Orange or Fine Gray. Candeleros and Teotihuacan figurines were also manufactured using local clays." Excavations confirmed this statement, and revealed that only three vessel forms were being manufactured. Santley raised an interesting hypothesis regarding Copaware (Copa Ware), which he believes was either made at Matacapan and traded to Teotihuacan, was a local imitation of the Teotihuacan Copoid Ware, or that the Copaware at Teotihuacan was inspired by the Matacapan product (1984:79-80, 1985:52). Copoid ceramics at Teotihuacan, made from local Teotihuacan Valley clays, were produced during the very late Early Tlamimilolpa phase through the early Metepec phase, ca 425-700 A.D. (Kolb 1987a, 1987b).

Nonetheless, the Teotihuacan to Matacapan "corridor" permitted the importation of tropical lowland products from the Gulf Coast into the Meseta Central Basin of Mexico and the dominant polity, Teotihuacan. In addition to cacao (probably drunk from copoid ceramic vessels), fruit, feathers, and semi-precious stones, I suggest that Matacapan would have been an ideal site involved in the Teotihuacan shell procurement system. Matacapan could have been a major collection center for Caribbean Marine Faunal Province mollusks, which would be transported by tumpline carriers to the Teotihuacan consumers. Thus far, Santley has not found any significant marine shell remains, but if such remains are found they could constitute a collecting and transshipment site (possibly a bodega or warehouse?), but, on the other hand, if such remains are not uncovered during the limited excavations, it might mean that the marine shell was all transshipped and the "bodega was bare" by the end of Classic times at Matacapan.

Chronologically, there would be a close "fit" between the Santa Maria Maquixco el Bajo TC-8 site and Matacapan, since the shell bodega at the former dates to Xolalpan times, the period of "heaviest" Teotihuacan influence and occupation at Matacapan. The importation of Pachuca green obsidian from the highland source into Matacapan further illustrated the close economic ties between the

Teotihuacan polity and Matacapan. The obsidian tools would have been useful in preparing cacao beans and tropical bird feathers for shipment into the Basin of Mexico.

The entire Gulf Coast from Laguna de Puebla Viejo (the locus of modern Tampico) in the north to Laguna de Terminos (Ciudad del Carmen) in the south was a source for cacao, other tropical products, and Caribbean marine shell valued by the Teotihuacan polity. Additional archaeological survey and excavation will be necessary in order to discern which of the six "Gulf Coast Routes" were used in the importation of coastal products and shell to the Teotihuacan Supra-Regional Center in the Basin of Mexico. The Gulf Coast Huastec and Classic Lowland Maya ceramics found at Teotihuacan, the El Tajin settlement in the "Teotihuacan Corridor," and Teotihuacan "influence" at Matacapan suggested the viability of "Gulf Coast Routes" A, B, C, and G; especially B (to El Tajin) and G (Matacapan). Depending on the route and variations because of "impassable" topography, route distances between the Gulf Coast and Teotihuacan ranged from 235-585 kilometers.

Pacific Coast Routes A-G

Commercial routes from Teotihuacan to west Mexico and the Pacific Coast are more difficult to identify because of "gaps" in the archaeological surveys. However, at least seven loci on the Pacific Coast could have provided cacao, tropical products, and marine shell to the Supra-Regional Center during the Classic Period. (See Figure 35.) I shall designate these as "Pacific Coast Routes" A through G. Precise evidence for some Pacific connections is often lacking (Bell 1971, 2974; Bernal 1965; Brand 1971; Chadwick 1971; Lister 1947, 1971; Meighan 1971).

"Pacific Coast Route A" proceeded northwest from the Teotihuacan Valley into the drainage of the Rio Tula of the Tula Region then westward overland through the present-day states of Queretaro and Guanajuato into the Rio Juchipila and Rio Grande de Santiago to the San Blas coast of Nayarit. Feldman (1974:232) recorded Spondylus at San Blas Site 8, Nayarit. The Early Ixtlan phase (ca 200 B.C. - 500 A.D.) in southern Nayarit, western Jalisco, and Colima was the time of shaft tomb construction (Furst 1966:94-97, 153-170; Mountjoy 1978:132) and the importation of Caribbean conchs used as shell trumpets. Strong interregional contacts were noted during the Los Cocos phase (ca 500-700 A.D.) but lesser contact with the Basin of Mexico (Mountjoy 1978:133-134). During the Postclassic, the San Blas region was important for its cacao (Feldman 1978:138).

FIGURE 35: PACIFIC COAST TO HIGHLANDS COMMERCIAL ROUTES DURING THE CLASSIC TEOTIHUACAN PERIOD

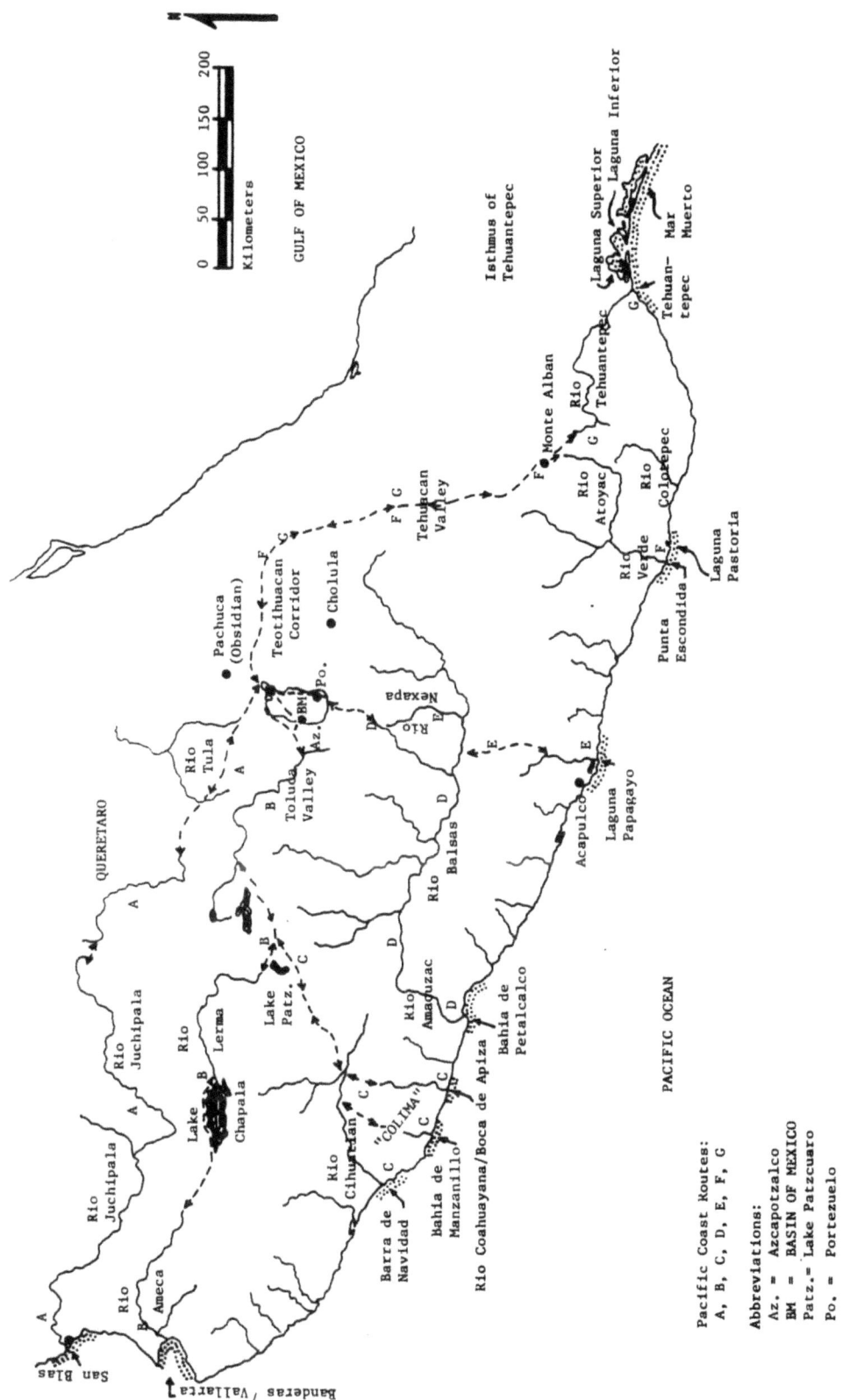

- 117 -

The West Mexican shaft tombs, ca 100-400 A.D., provided an ideological link to Teotihuacan, especially in the form of one Caribbean shell (Turbinella angulatus Sowerby) with Teotihuacan incised motif and fresco painting (Furst 1966:95, Pl. 45; Bell 1971:720, 748-749). No identifiable Classic Teotihuacan ceramics or architectural styles were noted, but the quantities of Caribbean conchs in several sites in northern and southern Nayarit and in the Lake Chapala region of Jalisco suggested a "wide trade network" (Bell 1971:720, 726, 728, 734; Kelley 1971:770-771). Radiocarbon dates ranged from 140 B.C. - 900 A.D., with the majority from 100-500 A.D.

Classic Teotihuacan "Thin Orange" pottery (apparently Alpha Type,) dated to 200-350 A.D., was recovered by Bell at the Cerro Encantado site on the Rio Verde in northern Jalisco near the border with Zacatecas. This site was located on overland routes to Zacatecas and Chalchihuites, which provided mineral resources to the Supra-Regional Center of Teotihuacan (ca 200-400 A.D.) (Millon 1981:227-228). Fragments of Caribbean-origin conch shell trumpets, noted at Cerro Encantado, also dated to the Classic Period.

"Pacific Coast Route B" passed through the Lower Teotihuacan Valley and skirted the northern shore of Lake Zumpango, or traversed the Basin of Mexico lakes via canoe. From the west side of the Basin, several routes led west into the Toluca Valley and into the drainage of the Rio Lerma, then proceeded along the shore of Lakes Patzcuaro and Chapala, to the Etzatlan Region into the drainage basin of the Rio Ameca and the coast. The Rio Ameca entered the Bahia de Banderas, the location of modern Puerto Vallarta, Jalisco. Weigand (1974:123, 127-130) identified a frescoed ware and Classic Teotihuacan ceramics, including Alpha "Thin Orange," at the Ahualulco site in the Etzatlan Region during the Ahualulco phase (ca 200-350 A.D.). During the early part of the subsequent Teuchtitlan phase (ca 350-900 A.D.) the region was "incorporated directly into the Teotihuacan sphere" of influence. A complex of sites at Teuchtitlan was involved in obsidian-working, and this resource probably piqued the interest of the Teotihuacan polity in the Etzatlan Region.

"Pacific Coast Route C" used the same routes from the Teotihuacan Valley into the Valley of Toluca and Rio Lerma drainage, but probably passed south of Lake Patzcuaro. This route continued overland into the drainages of either the Rio Armenaria to the Laguna de Cuyutlan and Bahia de Manzanillo in Colima, or followed the Rio Coahuayana to the Boca de Apiza area of Colima.

I know of no evidence suggesting Classic Teotihuacan contact via this potential route, but Spondylus were noted at Playa del Tesoro and Once Pueblos, Colima (Feldman 1974:233).

"Route D" followed the "Eastern Basin of Mexico Route (Route 1)" from Teotihuacan through the Texcoco Plain east of Portezuelo (Hicks and Nicholson 1964), a Teotihuacan Regional Center, and passed through the eastern sections of the Ixtapalapa, Chalco, and Xochimilco survey regions out of the Basin of Mexico into the tierra templada of Morelos (Parsons 1971, Sanders et al 1979, Parsons et al 1982, Kolb 1986:196). During the Classic Period in Morelos (ca 300-600 A.D.), there were 121 Teotihuacan-affiliated sites in the Rio Amatzinac Valley of eastern Morelos, while the Coatlan Region to the west had thirty-nine and the Valley of Xochicalco had thirteen (Hirth and Villaseñor 1981:140-143). The two former regions were probably part of the Teotihuacan "empire." Unfortunately not many marine shell specimens have been found, probably because of high soil acidity (Vaillant and Vaillant 1934:103). "Pacific Coast Route D" continued south through Morelos into the drainages of the Rio Amacuzac and especially the Rio Balsas (Litvak King 1978:119). The latter river emptied into the Bahia de Petalcalco. Sites in the middle Rio Balsas drainage had a few shell ornaments of Cardium and Conus, but no Spondylus or conchs were reported by Lister (1947:75), although Chadwick (1971:665, 676) noted conch shell trumpets, shell beads, bracelets and other ornaments at site V-42 (ca 700 A.D.) and shell necklaces at the Jiquilpan site (ca 500-600 A.D.), both in Michoacan. Feldman (1974:234) did not identify any Spondylus in Balsas Region sites. However, during the Late Postclassic, the town of Cacatulan, located at the mouth of the Rio Balsas, provided marine shells to the Aztec/Culhua Mexica (Barlow 1949:8-15).

"Pacific Coast Route E" from the Teotihuacan Supa-Regional Center followed the initial segments of "Route D" from the Basin of Mexico southward into Morelos and the Rio Nexapa drainage. The passage continued into the upper Rio Balsas then south through the Rio Papagayo system to Laguna Papagayo on the coast. Bahia de Acapulco was located immediately to the west. Spondylus inhabit this region of the coast of Guerrero, and Harvey (1971:631) stated that shell bracelets and beads were found "throughout Guerrero." During the Late Postclassic, this region provided cotton cloth, feathers, cacao, and seashells to the Aztec/Culhua Mexica (1971:603). "Route E" from Teotihuacan to the Acapulco region was the shortest distance to any Pacific coast Spondylus source, ca 325 km. The roles of the sites at Acapulco and La Sabana immediately to the east of

Acapulco are not known (Bernal 1951).

"Pacific Coast Route F" followed the initial segments of "Gulf Coast Route G" from Teotihuacan to the Tehuacan Valley and the vicinity of Tuxtepec (Tochtepec). Instead of continuing northeast to the Gulf Coast, "Route F" passed southeast via the Rio Quiotepec into the Valley of Oaxaca toward the Classic Period site of Monte Alban and the Rio Atoyac drainage. An overland passage south from the Rio Atoyac to the Rio Colotepec led to the Pacific coast at Punta Escondida (modern Puerto Angel). The Punta Escondida region has four sites, Sipilote and Puerto Angel on the coastal plain, and Pochutla and Tukidi in the river basin (Bernal 1965:790). The region was associated with the polities of the Valley of Oaxaca from Monte Alban I-V (ca 500 B.C. - 1500 A.D.) No specific evidence apparently exists to suggest that Teotihuacanos used this route.

"Route G" from Teotihuacan to the Isthmus of Tehuantepec followed "Route F" into the Valley of Oaxaca but entered the Rio Tehuantepec drainage basin to the southeast rather than turning due south into the Rio Atoyac and Rio Colotepec systems. The Rio Tehuantepec led to modern Salina Cruz on the Gulf of Tehuantepec. Immediately to the east were Laguna Superior, Laguna Inferior, and Mar Muerto where Spondylus were reported. The Gulf is a major commercial fisheries zone. Again, no specific Classic Teotihuacan contact can be documented.

The Pacific Coast from the Gulf of California to Guatemala was a potential Teotihuacan source for various tropical and non-tropical products, especially cacao and bird feathers, as well as Panamanian marine shell. The West Mexican ceramics found in the Teotihuacan Valley (Weigand 1972), and Classic Teotihuacan motifs and ceramics (especially Alpha "Thin Orange") found in West Mexican sites pointed to commercial and iconographic ties. Routes, A, B, D, and E appeared most viable on the basis of known archaeological and chronological data. "Route A" to the San Blas Region of Nayarit was significant as early as 200 B.C. for the importation of Caribbean conch shell trumpets, but Teotihuacan's presence was seen iconographically and perhaps ideologically. The Region potentially acted as a funnel for Teotihuacan's importation of raw materials from the north and northwest, including Chalchihuitean products. The presence of Teotihuacan ceramics and the obsidian workshops at Teuchtitlan in the Etzatlan Region provided a focus for Teotihuacan and "Route B" (Bahia de Banderas). Less certain evidence exists for "Pacific Coast Routes" D (Rio Balsas-Bahia de Petalcalco Region) and E (Rio Balsas-Bahia de Banderas). Less certain evidence exists for "Pacific Coast Routes" D (Rio

Balsas-Bahia de Petalcalco Region) and E (Rio Balsas-Bahia de Acapulco Region), the shortest distance to the tierra caliente and coastal zone. An added advantage associated with these latter two routes was the intermediate tierra templada of Morelos and its raw materials and products.

CHAPTER TEN

CLASSIC TEOTIHUACAN COMMERCE: THE ROLE OF MOLLUSKS

Introduction

Archaeological research on commercial systems requires, according to Earle and Ericson (1977:4), a four-step procedure: 1) characterization of the products and their sources (raw materials, marine shell, obsidian, and ideology); 2) descriptive modeling (the significance of the products and mechanics of distribution); 3) the application of ethnohistoric and ethnographic research (professional merchants, "ports of trade," commercial routes, and state controls); and 4) systemic modeling. The latter step is often conditioned, especially in Mesoamerican studies, by the cultural ecological approach and paradigms (Hirth 1984b:282-284). Following a review of the first three steps, I shall proceed to a systemic model to explain the waxing and waning of the Teotihuacan commercial networks.

The Shell Procurement System - The Teotihuacan Connection

In this monograph I have characterized the Classic Period Supra-Regional Center of Teotihuacan as a state and "empire" which had commercial ties to numerous regions of Mesoamerica. Teotihuacan's ability to control obsidian sources, production, and the distribution of finished products was the significant factor contributing to the waxing of Teotihuacan commerce, ca 100-650 A.D. The polity imported a variety of raw materials, foodstuffs, and luxury goods during the Classic Period, while exporting obsidian and religious ideology. Among the imports having ideological importance were marine mollusks, especially Spondylus and Strombus, seen in the wall murals and sculpture in structures and residences at the urban center. The manufacture of shell ornaments was only one of Teotihuacan's craft industries, but one which necessitated economic ties to both Gulf Coast and Pacific Coast sources.

I have postulated the existence of specialist merchants involved in long-distance trade. Some were "imperial" merchant-agents of the state, others were "private-sector" independent entrepreneurs perhaps licensed by the state. There is no evidence to suggest the existence of Classic Period "ports of trade," although these probably existed. Matacapan (Santley et al 1984, 1985) is a likely candidate or at least a collection point or node on a dendritic commercial network. Seven probable commercial routes from Teotihuacan to the Gulf Coast were identified, of which two (Routes B and G) to El Tajin and Matacapan, were most significant in the importation of tierra caliente products including tropical bird feathers, cacao, and Caribbean marine shell. An additional seven probable routes were suggested for the Teotihuacan-Pacific Coast network of which four (Routes A, B, D, and E) to San Blas, Etzatlan-Bahia de Banderas, Bahia de Petalcalco, and Bahia de Acapulco were significant. The San Blas and Etzatlan-Banderas routes provided access to Pacific Coast tierra caliente products and to important raw materials found to the north and northwest in Chalchihuites. Routes D and E also provided access to sources of cotton, cacao, and Panamanian Marine Faunal Province products.

Commerce in raw materials, finished products, and commodities in general would be accompanied by the diffusion of sociocultural and religious complexes, especially ideology and iconography (L. Parsons and Price 1971:169, L. Parsons 1978:25-28, Pasztory 1978:14-18). The commercial networks of Classic Teotihuacan were built upon the Middle Preclassic network developed by the Olmec (Flannery 1968:105), but were probably less elaborate than the extensive and intensive networks of the Aztec/Culhua Mexica (Barlow 1949; Bittman and Sullivan 1978; Berdan 1978, 1982; Sahagun 1954, 1959, 1961). Although Teotihuacan may have incorporated portions of the older distribution and exchange system, centralized state control was necessary because of the quantities of goods, materials, and foodstuffs involved and because of the spatial extent of the Classic networks (L. Parsons and Price 1971:181, Millon 1981:227-228). The quantities of "exotic," sumptuary commodities used within the Teotihuacan polity was not parallelled until the Late Postclassic. Sumptuary goods permit analogies to the Melanesian Kula Ring, an interaction sphere, or Northwest Coast potlatching (Flannery 1968:105, Shackleton and Renfrew 1970:1064-1065, L. Parsons and Price 1971:179-180, Paulsen 1974:604-605).

The evidence for Gulf Coast marine mollusk exploitation, directly or indirectly by Teotihuacan merchants or their agents, consisted of the shells themselves in ceremonial caches and burials in the urban center. Depictions of shells in lithic sculpture in Teotihuacan's ceremonial areas, especially on the Temple of Quetzalcoatl in the Ciudadela, and in wall mural art also provided evidence for the Gulf Coast connection. The depiction of shells and human footprints in the Palace of the Jaguars (Room 7b, Mural 4) reported by

Miller (1973:52, Figure 33) very likely denoted a commercial route. Likewise, the "shell diver" murals at the Tetitla residence (Portico 26, Murals 3 and 4) argued for knowledge by Teotihuacanos of the manner in which bivalves were obtained (Miller 1973:136, 1978:68-69). One may speculate that the inhabitants of the Palace of the Jaguars and Tetitla were high ranking pochteca or were the "corporate directors" of the shell procurement system. The Maya-style "old man" and Tepeu-like pigments in Room 7 Mural 5, and Maya-style seated figure in Room 27 at Tetitla (Miller 1973:134-136, 1978:68-69) suggested potential Maya merchants and/or shell procureres. The Caribbean Dentalium shell necklace in an ofrenda at Tetitla (Sejourne 1966c:165) was unique in that this was the only instance in which shells of this genus were found in any Classic or Postclassic Basin in Mexico sites. Quantities of marine shell fragments were noted in the fields adjacent to Tetitla, Yayahuala, and Zacuala as well as the Great Compound.

Such evidence, to me, argues that Tetitla was the residence of feather and marine shell merchant-procurers whose orientation was the Gulf Coast. I believe these merchants were "imperial" agents of the state and were Teotihuacanos rather than Maya. Cross-dating based on ceramics, mural style, and architectural features at Tetitla would suggest a chronology of 300-650 A.D., consistent with Matacapan (Santley et al 1984, 1985).

Teotihuacan commercial connections with the Pacific Coast were evidenced by the shells themselves and depictions in lithic sculpture, potentially including Spondylus calcifer on the Temple of Quetzalcoatl. The presence of Caribbean Strombus and/or Turbinella and Panamanian Spondylus could connote religious-ideological, iconographic, and economic ties to regions of both the Gulf and Pacific coasts.

The extraordinary quantity of Spondylus calcifer in the Room 2 bodega at the Santa Maria Maquixco el Bajo (TC-8:3) suggested that inhabitants at that site were potentially directly involved as shell procurement merchants, or, more likely, provided a central warehouse (bodega or petalcalco) for "imperial" and/or "private sector" merchants who travelled to the Pacific Coast. Maquixco was the largest site west of the urban center in the Teotihuacan Valley and would have direct access to the northern Basin of Mexico, the Toluca Valley, and commercial routes to West Mexico. The TC-8 site was 5.1 kilometers west of the Great Compound, Teotihuacan's primary marketplace, and 4.9 kilometers southwest of the Spondylus shell-working area in Oztoyahualco (Millon 1973:38-39, 1981:227), while the Tlamimilolpa phase shell workshop at Tlajinga 33 was 6.5 kilometers to the southeast (Storey and Widmer 1982:39).

The unique location of TC-8:3 further suggested that its inhabitants may have functioned as short-distance suppliers to the market(s) and workshop(s) of the urban center but were not directly involved in procuring the shell from Pacific Coast sources. The Spondylus calcifer in the two TC-8:3 ofrendas and West Mexican sherds (Chatchihuites and other Classic Period wares from Jalisco, Nayarit, and Zacatecas) were dated ca 250-500 A.D. Only a few shells of Caribbean origin were recovered at TC-8, but Tajin Huastec III sherds were also found, suggesting some contact with, and knowledge of, the Gulf Coast as well. Lowland Classic Maya sherds, dated ca 300-600 A.D. at Altar de Sacrificios and Tikal, were also recovered at TC-8:3, arguing for indirect contact with the Yucatan, perhaps via the Gulf Coast and Matacapan.

The Shell Procurement System - A Commercial Model

A model derived from cultural and economic geography (Kebble 1967, Vance 1971) and Maya commerce (Rathje et al 1978) may be applicable to Teotihuacan's importation of marine shell from the Gulf and Pacific provinces and the waxing and waning of the Teotihuacan polity and its commerce. Keeble (1967:257-268) elaborated "regional income inequality" and "export-based" paradigms in considering national, state-level economic development. In the former, economic development was not evenly spread over the entire area of a political unit (the Teotihuacan polity), but was concentrated at several points (Teotihuacan, Portezuelo, Azcapotzalco, Tepeapulco, and perhaps Matacapan), producing a mosaic of regions at different levels of economic prosperity. Once economic growth began with Teotihuacan's control of the obsidian industry, it set into motion a series of feedback loops resulting in the accelerated development of economics which required a minimum demand, but also produced effects that stimulated further growth. This would be a "cumulative causation" process (Keeble 1967:258-259).

A critical economic threshhold was crossed as imports greatly exceeded exports. Teotihuacan imported a vast array of raw materials, finished products, "exotic" goods, and foodstuffs which far surpassed the quantities of obsidian exports. By about 650 A.D. the Teotihuacan Valley could provide only enough food to feed approximately 79,000 persons of the estimated 125-250,000 urban residents, let alone the 21,800-22,600 rural inhabitants in the Valley (Kolb 1979a:296-298, Table 60;

Sanders et al 1979:393-395; Hirth 1984a:3). Therefore, food importation became a significant concern of the state. The exportation of obsidian and importation of raw materials, foodstuffs, etc. had originally caused a regional multiplier effect (Keeble 1967:275) in which the production, employment and income of one region stimulated the expansion of other groups in other regions (Cholula, Xochicalco, Monte Alban, etc.).

A polarization or "backwash" effect (1967:259) was seen as urban Teotihuacan's growth attracted labor, capital and commodities from the surrounding regions in the Meseta Central. The process of "centrifugal spread" counteracted the "backwash" effect because the surrounding regions increased the demand for Teotihuacan's finished products and obsidian (Keeble 1967:259-260). If the spread increased sufficiently beyond the backwash effect, new growth centers (Cholula and Xochicalco) would be established or enhanced at the expense of the Teotihuacan polity. This appears to have begun ca 650 A.D., and Teotihuacan commerce waned.

In order to correct the import-export imbalance, "imperial" Teotihuacan merchants specializing in the exportaiton of a small number of products needed to expand their geographic horizons, while those who dealt in a wide variety of products needed to restrict their realms (Vance 1971:65), but apparently did not. The distant entrepôts were abandoned, with the resulting decline in the importation of tropical products and marine shell, as the state concentrated on the more immediate subsistence needs. Hence, the importation of luxury goods and exotic materials associated with ideology and status enhancement declined. Teotihuacan's commercial network fell into disrepair, and new emerging polities of the early Postclassic (Cholula, Xochicalco, and Tula) appropriated segments for their own economic purposes.

During the Metepec phase (ca 650-750 A.D.) armed conflict was apparently resorted to by the Teotihuacan state as it sought to reduce the control by the emerging polities of food and other imports. A series of internal and external difficulties, many accumulating throughout the Late Classic, ultimately led to the collapse of the Teotihuacan "empire" and state, and the "destruction" of the Supra-Regional Center by 750 A.D. (Millon 1981:235-238). The military was prominently represented at Teotihuacan during the Metepec phase, and, as Millon stated, "this may be both a symptom of difficulty and a cause of it" (1981:236).

INDEX TO THE APPENDICES

I. Specific Mollusks Represented in Classic Period Teotihuacan Valley Rural Sites: Site, Genus and species

II. Summary of Specific Mollusks Represented in Classic Period Teotihuacan Valley Rural Sites: Classes and Marine Faunal Provinces

III. Comments on Selected Mollusks Represented in Classic Period Teotihuacan Valley Rural Sites

IV. Mollusks Represented in Classic Period Basin of Mexico Sites

V. Marine Shell in Classic Period Sites in the Teotihuacan Valley: The Sites.

Abbreviations:

CLASS

- G Gastropoda
- P Pelecypoda
- G/P Gastropoda or Pelecypoda

RANGE

- F   Freshwater Riverine mollusks
- C   Caribbean Marine Faunal Province
- P   Panamanian Marine Faunal Province
- C/P Caribbean or Panamanian Marine Faunal Province

APPENDIX I

SPECIFIC MOLLUSKS REPRESENTED IN TEOTIHUACAN VALLEY RURAL SITES

MARINE AND RIVERINE MOLLUSKS

TAXON — Genus species	Astraea prob. A. undosa Wood	Cerithidea scalariformis Say	Chama buddiana C. F. Adams	Chama echinata	Fasciolaria princeps Sowerby	Fissurella angusta Gmelin	Glycymeris prob. G. subobsoleta Carpenter	Lampsilis discus Lea	Latirus ceratus Wood	Melongena prob. M. patula Broderip & Sowerby	Morum oniscus Linne	Oleacinidae (Fam.), Euglandina (Gen.)	Olivia (?)	Oliva sayana Ravenel or O. porphyria Lamarck	Spondylus calcifer Carpenter	Unionidae (Fam.), Elliptio (Gen.) Unio discus	Vermetidae (Fam.), Vermetus (Gen.)	Unidentified	Unidentified	TOTAL
Class:	G	G	P	P	G	G	P	P	G	G	G	P	G	G	P	P	G	G/P	G/P	
Range:	P	C	P	P	C	C	P	F	P	P	C	F	C/P	C/P	P	F	P	C/P	C/P	
TC-8:Pyramid	0	0	0	0	0	0		0	0	0	0	0	0		0	0	0	0	0	0
TC-8:1-2	1	1	0	0	0	0	0	0	3	1	1	5	1+		194+	3+	X	4	3+++	215
TC-8:3	0	0	33	0	2	0	0	1	0	0	0	0	0		3862+	6+	X	20	1+	3927
TC-8:4	0	0	0	0	1	0	0	0	0	0	0	0	0		28	0	X	0	0	29
TC-8:*	0	0	16	0	0	0	0	0	0	0	0	0	0		0	0	0	1	0	17
SUBTOTAL	1	1	49	0	3	0	0	1	3	1	1	5	1	0	4084	9	X	25	4	4188
TC-10:2						1								1		0+				2
TC-49:1-3	0	0	0	0	0	0	0	0	0	0	0	0	0	0	0	1	0	0	1	2
SUBTOTAL	0	0	0	0	0	1	0	0	0	0	0	0	0	1	0	1	0	0	0	4
EXCAVATED TOTAL	1	1	49	0	3	1	0	1	3	1	1	5	1	1	4084	10	X	2	4	4192

SITE EXCAVATIONS

Class: Gastropoda — Pelecypoda

Range: Fresh Water — Caribbean — Panamanian

+ Worked Specimens

* Obliterated Number

X Present

TAXON	Astraea prob. A. undosa Wood	Cerithidea scalariformis Say	Chama buddiana C. F. Adams	Chama echinata	Fasciolaria princeps Sowerby	Fissurella angusta Gmelin	Glycymeris prob. G. subobsoleta Carpenter	Lampsilis discus Lea	Latirus ceratus Wood	Melongena prob. M. patula Broderip & Sowerby	Morum oniscus Linne	Oleacinidae (Fam.), Euglandina (Gen.)	Olivia (?)	Oliva sayana Ravenel or O. porphyria Lamarck	Spondylus calcifer Carpenter	Unionidae (Fam.), Elliptio (Gen.) Unio discus	Vermetidae (Fam.), Vermetus (Gen.)	Unidentified G/P C/P	Unidentified G/P C/P	TOTAL
Genus species Class: Gastropoda / Pelecypoda Range: Fresh Water / Caribbean / Panamanian + Worked Specimens ☆ Obliterated Number x Present	G / G	G / G / C	P / C / P	P / P / P	G / G / C	G / C / C	P / P / C / P	P / F / P	G / P / P	G / P / P	G / C / C	P / F / F	G / C / P	G / C / P	P / P / P	P / F / P	G / P / P	G/P C/P	G/P C/P	
SITE EXCAVATIONS Class: / Range:	P		P	P			P								P					
SITE SURVEYS																				
TC-2:1	1			0			0											0	2++	2
TC-8:14				0			0											1	0	1
TC-8:28				0			1											0	0	1
TC-13:3				1			0											0	0	1
TC-40:7				0			0											1	0	1
TC-73:36				1			0											0	0	1
TC-91:1				0			1											0	0	1
SURVEY TOTAL				2			2											2	2	8
GRAND TOTAL	1	49	2	3	1	2	3	1	1	5	1	1			4084	10	X	4	6	4200

- 129 -

APPENDIX II

SUMMARY OF SPECIFIC MOLLUSKS

REPRESENTED IN CLASSIC PERIOD TEOTIHUACAN VALLEY RURAL SITES

SITE	Total	CLASS			RANGE			
		G	P	G/P	F	C	P	C/P
TC-8:Pyramid	0	0	0	0	0	0	0	0
TC-8:1-2	215	6	202	7	8	0	199	8
TC-8:3	3927	4	3902	21	7	4	3895	21
TC-8:4	29	1	28	0	0	1	28	0
TC-8*	17	0	16	1	0	0	16	1
TOTAL TC-8 EXC.	4188	11	4148	29	15	5	4138	30
TC-10:2	3	2	0	1	0	1	0	2
TC-49:1-3	1	0	1	0	1	0	0	0
TOTAL MINOR EXC.	4	2	1	1	1	1	0	2
EXCAVATED TOTAL	4192	13	4149	30	16	6	4138	32
TC-2:1	2	0	0	2	0	0	0	2
TC-8:14	1	0	0	1	0	0	0	1
TC-8:28	1	0	1	0	0	0	1	0
TC-13:3	1	0	1	0	0	0	1	0
TC-40:7	1	0	0	1	0	0	0	1
TC-73:36	1	0	1	0	0	0	1	0
TC-91:1	1	0	1	0	0	0	1	0
SURVEY TOTAL	8	0	4	4	0	0	4	4
GRAND TOTAL	4200	13	4153	34	16	6	4142	36

APPENDIX III

SELECTED MOLLUSKS REPRESENTED IN TEOTIHUACAN VALLEY RURAL SITES

The mollusks reported below were recovered from survey and excavations (TC-8:1-2, 3, 4; TC-10:B; TC-49:1-3). The unworked Spondylus calcifer Carpenter, 1857 shells from TC-8 are not listed.

	Site and Excavation Number	Family, Genus species	Description
1	TC-2:1 (Survey)	Unidentified	Worked, abraided marine shell fragment, possibly from a shell "cup," ornament or atl atl finger loop.
2	TC-2:1 (2) (Survey)	Unidentified	Worked fragment of marine shell with drilled hole (3.5 mm diameter), possibly an ornament.
3	TC-8:1-2 (1502) (Excavation)	Spondylus calcifer Carpenter, 1857	Tubular bead from a "Pacific Thorny Oyster," Panamanian Marine Faunal Province.
4	TC-8:1-2 (1561) (Excavation)	Unidentified	Worked, cut and abraided marine shell fragment.
5	TC-8:1-2 (1708) (Excavation)	Unidentified	Worked, abraided conical-shaped marine shell ornament fragment.
6	TC-8:1-2 (1737) (Excavation)	Unidentified	Worked, concave disk-like marine shell ornament.
7	TC-8:3 (8114) (Excavation)	Spondylus calcifer Carpenter, 1857	Tubular bead from a "Pacific Thorny Oyster," Panamanian Marine Faunal Province.
8	TC-8:3 (8169) (Excavation)	Unidentified	Worked, abraided and drilled disk-like marine shell ornament (pendant?).
9	TC-8:3 (8321) (Excavation ofrenda)	Unio discus Lea, 1838	Worked, complete artifact (trilobed bird wing or "paw-wing" of a "Freshwater Clam" from a riverine context or freshwater Basin of Mexico lake.
10	TC-8:14 (Survey)	Unidentified	Worked fragment of marine shell, possibly an ornament.
11	TC-8:28 (Survey)	Glycymeris subabsolata Carpenter	Unworked fragment of a "Bittersweet Clam" from the Panamanian Marine Faunal Province.
12	TC-10:2 (10,695) (Excavation)	Unidentified	Burned fragment of marine shell.
13	TC-10:2 (10,713) (Excavation)	Fissurella angusta Gmelin, 1791	Unworked complete shell of a "Keyhole Limpet" from the Caribbean Marine Faunal Province.

14	TC-10:2 (10,761) (Excavation)	*Oliva sayana* Ravenel, 1834 OR *Oliva (Oliva) porphyria* Linnaeus, 1758	Unworked, burned fragments of a "Lettered Olive" from the Caribbean Marine Faunal Province, or, less likely, a "Tent Olive" from the Panamanian Marine Faunal Province.
15	TC-13:3 (Survey)	*Chama echinata* Broderip, 1835	Unworked fragment of a "Jewel Box" from the Panamanian Marine Faunal Province.
16	TC-40:7 (Survey)	Unidentified	Unworked fragment of marine shell.
17	TC-49:1 (Tenango Excavation)	Unidentified	Unworked fragment.
18	TC-49:3 (10,575) (Tenango, Excavation)	Unionidae Family, *Elliptio* genus	Worked fragment of a "Freshwater Clam" from a riverine context or freshwater Basin of Mexico lake.
19	TC-73:36 (Survey)	*Chama echinata* Broderip, 1835	Unworked fragment of a "Jewel Box" from the Panamanian Marine Faunal Province.
20	TC-91:1 (Survey)	*Glycymeris subobsolata* Carpenter	Unworked fragment of an immature (non-adult) "Bittersweet Clam" from the Panamanian Marine Faunal Province.

APPENDIX IV

MOLLUSKS REPRESENTED IN BASIN OF MEXICO SITES

Key to Appendix:

Abbreviations:

- Range:
  - F Freshwater (riverine mollusks)
  - CM Central Mexico (land snails)
  - C Caribbean Marine Faunal Province
  - P Panamanian (Panamic or Pacific) Marine Faunal Province
  - P/C Panamanian or Caribbean Marine Faunal Provinces
  - BC Baja California subprovince of Panamanian Marine Faunal Province

- Other:
  - T. Templo
  - Teo. Teotihuacan
  - TC- Teotihuacan Valley Project, Teotihuacan (Period) Classic (site)
  - * Mollusks reported but not classified or reported as unclassified

Sites and References: Classic Period

La Ciudadela, Teotihuacan (Gamio 1922a:187-190)

Templo de Agricultura, Teotihuacan (Rubin de la Borbolla 1947:64-65)

Templo de Quetzalcoatl, Teotihuacan (Rubin de la Borbolla 1947:64-65)

Palacio de Quetzalpapalotl, Teotihuacan (Acosta 1964)

Xolalpan, Teotihuacan (Linne 1934:138-159)

*Tlajinga 33, Teotihuacan (Storey and Widmer 1982:80-97)

Tlamimilolpa, Teotihuacan (Linne 1942:150-153)

Tetitla, Teotihuacan (Sejourne 1966a, 1966b)

Yayahuala, Teotihuacan (Sejourne 1966b)

Zacuala, Teotihuacan (Sejourne 1959, 1966b)

Rio San Juan Tlatel at Highway (Noguera 1955)

Museo de Teotihuacan (Gamio 1922a:187-190)

Teotihuacan "art" (Cook de Leonard 1971:216)

TC-sites (Kolb 1973a)

Sites and References: Preclassic and Postclassic Periods

    El Arbolillo (Vaillant 1935a:249-250, Pires Ferreira 1978)

    Ticoman (Vaillant 1931:312-313, Richards and Boekelman 1937)

    Zacatenco (Vaillant 1930:49-50, Pires-Ferreira 1978)

    *Tlapacoya (Barba de Piña Chan 1956)

    *Azcapotzalco (Vaillant 1934)

    *Chiconautla (Vaillant 1935b)

    "Las Palmas," San Francisco Mazapan, Teotihuacan (Linne 1934:158)

    Portezuelo (Feldman n.d.)

    El Risco (Mayer-Oakes 1959)

    *Santiago Ahuitzotla (Tozzer 1921:42)

Chronology:

Preclassic Sites:	El Arbolillo, El Arbolillo West, Ticoman, Tlapacoya, Zacatenco
Classic Sites:	Urban Teotihuacan temples and residences, TC-site series, El Risco, Portezuelo, Tlapacoya
Classic Phases:	
Miccaotli Phase:	La Ciudadela, Templo de Agricultura, Templo de Quetzalcoatl, Tlapacoya
Tlamimilolpa Phase:	Tlamimilolpa, Tlajinga 33, Palacio de Quetzalpapalotl
Xolalpan Phase:	Xolalpan, Palacio de Quetzalpapalotl, Tetitla, Yayahuala, Zacuala, Rio San Juan at Highway, TC-8, Tlajinga 33
Postclassic Sites:	Azcapotzalco, Chiconautla, "Las Palmas," El Risco, Santiago Ahuitzotla, Tlapacoya

MOLLUSKS REPRESENTED IN BASIN OF MEXICO SITES
SUMMARY

Identified Family/Genera/species: n = 63

Chronology:	Preclassic	Preclassic to Classic	Classic	Classic to Postclassic	Postclassic	Total
Range:						
C	6	1	16	0	0	23
P	3	1	19	1	2	26
P/C	0	1	4	0	0	5
BC	0	0	4	0	0	4
CM	0	1	1	0	0	2
F	1	0	1	0	0	2
F(C)	0	0	1	0	0	1
	10	4	46	1	2	63

Entries (Sites, Museums, "Art," etc.) n = 111

Chronology:	Preclassic	Classic	Postclassic	Total
C	7	22	0	29
P	6	37	3	44
P/C	2	14	0	14
BC	0	6	0	6
CM	1	3	0	4
F	1	6	0	6
F(C)	0	3	0	3
	17	91	3	111

MOLLUSKS REPRESENTED IN BASIN OF MEXICO SITES

Genus and species	Range	Site (Chronology)
Arca grandis Broderip	P	"Las Palmas" Teo. (Postclassic)
Arca pextata	C	Ticoman (Preclassic)
Arca ponderosa	C	Ticoman (Preclassic)
Argopecten circularis Sowerby, 1835	P	Tlamimilolpa, Teo. (Classic)
	P	Museo de Teotihuacan (Classic)
Astraea undosa Wood, 1828	BC/P	TC-8 (Classic)
Atrina maura Sowerby, 1935	P	Tlamimilolpa, Teo. (Classic)
Busycon perversum Linnaeus	C	Tlajinga 33, Teo. (Classic)
Cardium (Trachycardium) isocardia Linnaeus, 1758	C	Tlamimilolpa, Teo. (Classic)
Cassis (genus) Scopoli, 1777	C	El Arbolillo (Preclassic)
Cerithidea scalariformis Say, 1822	C	TC-8 (Classic)
Chama (genus) Linnaeus, 1758	P/C	Rio San Juan/Highway, Teo. (Classic)
Chama buddiana C. B. Adams, 1852	P	TC-8 (Classic)
Chama coralloides Reeve, 1846	P	TC-13 (Classic)
	P	TC-73 (Classic)
	P	"Las Palmas," Teo. (Postclassic)
	P	Tlamimilolpa, Teo. (Classic)
Chama echinata	P	TC-13
Conus interruptus Broderip and Sowerby, 1829	P	Tlamimilolpa, Teo. (Classic)
Dentalium (genus) Linnaeus, 1758	C	Ticoman (Preclassic)
Dentalium (Graptacme) semipolitum Broderip and Sowerby, 1829	P	Tetitla, Teo. (Classic)
Fasciolaria (genus) Lamarck, 1799	P/C	La Ciudadela, Teo. (Classic)
	P/C	Tetitla, Teo. (Classic)
	P/C	Zacuala, Teo. (Classic)

Fasciolaria (Pleuroploca) gigantea Kiener, 1840	C	T. de Agricultura, Teo. (Classic)
	C	T. de Quetzalcoatl, Teo. (Classic)
	C	Tlamimilolpa, Teo. (Classic)
	C	TC-8 (?)
Fasciolaria (Pleuroploca) gigantea Kiener, 1840	C	Museo de Teotihuacan (Classic)
	C	Museo de Teotihuacan (Postclassic)
	C	Teotihuacan "art" (Classic)
	C	TC-8 (?)
Fasciolaria (Pleuroploca) princeps Sowerby, 1825	P	Xolalpan, Teo. (Classic)
	P	Teotihuacan "art" (Classic)
Fissurella angusta Gmelin, 1791	C	TC-10 (Classic)
Fusus dupetitthouarsi Kiener, 1840	P	"Las Palmas," Teo. (Postclassic)
Glycymeris subobsolata Carpenter	BC/P	TC-8 (Classic)
	BC/P	TC-91 (Classic-Postclassic)
Lampsilis discus Lea, 1838	F(C)	TC-8 (Classic)
	F(C)	Tlamimilolpa, Teo. (Classic)
	F(C)	Cerro Portezuelo (Classic)
Latirus ceratus Wood, 1828	P	Xolalpan, Teo. (Classic)
	P	TC-8 (Classic)
Limneae columella	CM	Ticoman (Preclassic)
Lithophaga (Myoforceps) aristata Dillwyn, 1817	P	Tlamimilolpa, Teo. (Classic)
Lyropecten (Nodipecten) subnodsa Sowerby, 1835	P	Xolalpan, Teo. (Classic)
	P	Museo de Teotihuacan (Classic)
Meleagrina margaritacea Lamarck, 1819	P	Xolalpan, Teo. (Classic)
	P	Ticoman (Preclassic)
Melongena patula Broderip and Sowerby, 1829	P	El Arbolillo West (Preclassic)
	P	TC-8 (Classic)
	P	Xolalpan, Teo. (Classic)

Morum oniscus Linnaeus, 1767	C	TC-8 (Classic)
Natica (Stigmaulax) broderipiana Recluz, 1844	P	Xolalpan, Teo. Classic)
Neritina picta Sowerby, 1832	P	Zacatenco (Preclassic)
Neritina reclivata Say, 1822	C	Ticoman (Preclassic)
	C	El Arbolillo (Preclassic)
	C	El Arbolillo West (Preclassic)
Oleacinidae (family), Euglandina (genus), Oleacina (species)	CM	TC-8 (Classic)
	CM	Ticoman (Preclassic)
Oleacina coronata	CM	Ticoman (Preclassic)
Oliva (genus) Bruguierre, 1789	P/C	TC-8 (Classic)
	P/C	Tetitla, Teo. (Classic)
	P/C	Yayahuala, Teo. (Classic)
	P/C	Zacuala, Teo. (Classic)
Oliva (Oliva) angulata Lamarck, 1811	BC/P	T. de Quetzalcoatl, Teo. (Classic)
Oliva (Oliva) porphyria Linnaeus, 1758	P	TC-10 (Classic-Preclassic)
	P	La Ciudadela, Teo. (Classic)
	P	T. de Quetzalcoatl, Teo. (Classic)
	P	Teotihuacan "art" (Classic)
Oliva (Oliva) reticularis Lamarck, 1811	C	Tlamimilolpa, Teo. (Classic)
	C	Ticoman (Preclassic)
Oliva (Oliva) sayana Ravenel, 1834	C	TC-10 (Classic)
Olivella (genus) Swainson, 1840	P/C	Museo de Teotihuacan (Classic)
Olivella matica Say, 1840	C	Tlajinga 33, Teo. (Classic)
Orthalicus zebra (land snail)	CM	El Arbolillo (Preclassic)
	CM	Zacatenco (Preclassic)

Pectinidae (family), Pecten (genus) Muller, 1776	P	T. de Quetzalcoatl, Teo. (Classic)
	P	Tetitla, Teo. (Classic)
	P	Museo de Teotihuacan (Classic)
Pecten turgidus Lamarck	C	Museo de Teotihuacan (Classic)
	C	Teotihuacan "art" (Classic)
Pecten jacobaeus Linnaeus	P	Museo de Teotihuacan (Classic)
Pecten nodsus Lamarck	BC/P	Museo de Teotihuacan (Classic)
	BC/P	Teotihuacan "art" (Classic)
Pteria margaritifera Linnaeus, 1758	P	Tlamimilolpa, Teo. (Classic)
Quadula	F	El Arbolillo (Preclassic)
Spondylus (genus) Linnaeus, 1758	P/C	Oztoyahualco, Teo. (Classic)
	P/C	Quetzalpapalotl, Teo. (Classic)
	P/C	Tetitla, Teo. (Classic)
	P/C	Yayahuala, Teo. (Classic)
	P/C	Zacuala, Teo. (Classic)
	P/C	Zacatenco (Preclassic)
Spondylus americanus Gmelin, 1781	C	Tlamimilolpa, Teo. (Classic)
Spondylus calcifer Carpenter, 1857	P	TC-8 (Classic)
	P	T. de Quetzalcoatl, Teo. (Classic)
	P	Tlajinga 33, Teo. (Classic)
Spondylus princeps Broderip, 1833	P	"Las Palmas," Teo. (Postclassic)
	P	Tlamimilolpa, Teo. (Classic)
	P	Xolalpan, Teo. (Classic)
Strombus (Tricornis) galeatus Wood, 1828	P	Museo de Teotihuacan (Classic)
	P	Teotihuacan "art" (Classic)
Strombus gigas Linnaeus, 1758	C	Rio San Juan/Highway, Teo. (Classic)
Strombus pugilis Linnaeus, 1758	C	Tlamimilolpa, Teo. (Classic)

Thais deltoidea Lamarck, 1822	C	Tlamimilolpa, Teo. (Classic)
Thais fasciata Reeve	C	Tlamimilolpa, Teo. (Classic)
Turbinella scolymus Gmelin	C	Tlamimilolpa, Teo. (Classic)
	C	Xolalpan, Teo. (Classic)
	C	Museo de Teotihuacan (Classic)
	C	Teotihuacan "art" (Classic)
Turritella leucostoma Valenciennes, 1832	P	Tlamimilolpa, Teo. (Classic)
Unionidae (family), Elliptio (genus)	F	TC-8 (Classic)
	F	TC-49 (Classic)
	F	Tetitla, Teo. (Classic)
	F	Yayahuala, Teo. (Classic)
Unio discus Lea, 1838	F	Tlamimilolpa, Teo. (Classic)
	F	TC-8 (Classic)
Vermetidae (family), Vermetus (genus) Daudin, 1800	P	TC-8 (Classic)

APPENDIX V

MARINE SHELL IN CLASSIC PERIOD SITES IN THE TEOTIHUACAN VALLEY: THE SITES

This Appendix provides descriptive information on the eight Teotihuacan Valley Classic Period sites (TC-2, 8, 10, 13, 40, 49, 73, 91) in which shell was recovered during archaeological reconnaissance and/or excavation (1960-1965). The data presented is summarized from Kolb (1979a), and includes information on the sites' Background (proper noun name, nature of the artifact collections, published reports, etc.), Natural Setting (ecological location, altitude, soil types, erosion, vegetation, etc.), Modern Utilization (current use, modern structures, crops planted, etc.), Archaeological Remains (descriptions of each mound or tlatel composing the site, size in square meters, degree of erosion or disturbance, and archaeological periods and phases represented in the artifact collections, etc.), and Classification (settlement pattern type and subtype).

The following terms are used to characterize the artifact collections (sherds, complete vessels, figurines and other artifacts of known chronological phase):

Terms Used	Percentage (%)
Absent.	0
Traces.	1 - 5
Sparse.	6 - 10
Sparse to Moderate.	11 - 25
Moderate.	16 - 50
Moderate to Heavy	51 - 75
Heavy	76 - 100

Abbreviations used:

TF. . . . . . . Teotihuacan (Valley) Formative (site) or (Survey Team)

TC. . . . . . . Teotihuacan (Valley) Classic (site) or (Survey Team)

TT. . . . . . . Teotihuacan (Valley) Toltec (site) or (Survey Team)

TA. . . . . . . Teotihuacan (Valley) Aztec (site) or (Survey Team)

TC-2

Background. Ths site, also known as Techachal-Tetechalchal (El Molino Viejo) de Atlatongo, was surveyed by Kolb in 1963 and resurveyed by Kolb in 1972. The artifact collections from the Classic survey were processed by Kolb in 1963 and 1964 and consisted of two samples. Collections made by the TF, TT, and TA survey teams were also studied by Kolb. The ceramic materials from the Classic survey were excellent for chronological purposes, while figurine data were excellent and the Classic field reports were of high quality. Published and unpublished materials on this site include Marino (1965) and Sanders (1965).

Natural Setting. Site TC-2 is located in the Middle Valley, North Piedmont, of the Teotihuacan Valley between 2,280-2,350 m in the Lower Piedmont Ecological Zone. Soils in the site area have a sandy to loamy texture and are tan in color with depths ranging from 0 to at least 35 cm. There is moderate erosion in the northeastern area, and tepetate is exposed along the southeast area of the site. Heavy concentrations of rock and tezontle fragments are found in the site area. Vegetation in the vicinity includes pirul trees, maguey, and grasses of various types. Other natural features include a wash-road to the northwest and another wash-road to the southeast.

Modern Utilization. The cultural features include no structures and no jagueys. The Canal de San Antonio is northeast of the site location. The site area is used for agricultural purposes including the cultivation of maize, barley, wheat, beans, and nopal (cactus). Maguey are also found in the area.

Archaeological Remains. The total multicomponent site occupies 6.3 ha, while the Classic component also occupies 6.3 ha. Two mounds were identified of which both have Classic occupation. The site has three phases of Preclassic, two phases of Classic, and three phases of the Post Classic represented for a total of eight. Associated non-Classic sites include TF-186 and TF-6, TT-12, TT-13, and TA-158. While Classic sites TC-4 and TC-18 are located to the south and TC-5 to the north. The following mounds were identified:

> TC-2:1, 3,915 m^2, somewhat eroded and damaged. Artifactual remains indicated sparse Patlachique, sparse Early Tzacualli, sparse Miccaotli, sparse Early Tlamimilolpa, moderate Late Tlamimilolpa, sparse Early Xolalpan, sparse Xometla, sparse Mazapan, and sparse Aztec components. Two

unidentified, worked shell fragments were recovered. One, a worked and abraided piece, possibly came from a "cup" or an ornament, but may have been the finger loop of an <u>atl atl</u> (the throwing board of a spear thrower), while the second, which had a drilled hole (3.5 mm in diameter), probably was an ornament fragment. (See Appendices I, II and III.)

TC-2:2, 4,400 m², eroded and slightly damaged. Artifactual remains indicated sparse Early Tzacualli, sparse Early Tlamimilolpa, sparse Late Tlamimilolpa, sparse Early Xolalpan, sparse Xometla, sparse Mazapan, and heavy to moderate Aztec components.

The site had a moderate distribution of lithic materials including obsidian blades, cores, projectile points, and especially scrapers. Ground stone tools, especially manos and metates, were also noted. Other special lithic materials included basalt and quartzite fragments.

<u>Classification</u>. TC-2 was interpreted by Marino (1965:163) as a small compact hamlet or village dating to mid- to late Teotihuacan, and by Sanders (1965:10-8, 109) as a village. The site is reclassified as a Small Nucleated Village with a probable population of about 200 individuals (maximum 400) during Tlamimilolpa to Xolalpan times. The site probably progressed from a Hamlet during the Early Classic to a Small Nucleated Village during the Middle Classic and was apparently unoccupied during the Late and Terminal Classic.

TC-8

Background. This site, also known as Santa Maria Maquixco el Bajo or Los Tres Reyes Yzquitlan or Loma de las Calaveras, was surveyed by Kolb in 1963 and resurveyed by Kolb in 1970 and 1972. The artifact collections from the Classic survey were processed by Kolb from 1963 to 1964 and consisted of 74 collections plus a large sample from four excavated sites. No collections were made by other survey teams. The ceramic materials from the Classic survey were excellent for chronological purposes, while figurine data were excellent and the Classic field reports were of excellent quality. Published and unpublished material on this site include Marino (1965, 1975) and Sanders (1965, 1966).

Natural Setting. Site TC-8 is located in the Lower Valley, North Piedmont, of the Teotihuacan Valley between 2,320 m in the Lower Piedmont Ecological Zone. Soils in the site area have a sandy to loamy texture and are tan to light brown in color with depths ranging from 0 to at least 130 cm. There is moderate erosion in the eastern site area, and tepetate is exposed along the eastern side of the site. In addition, there is serious erosion to the southern portion of the site where several washes have formed. A canalized barranca is found at the eastern side of the site and areas of serious sheet erosion to the north of the site area. Vegetation in the vicinity includes nopal, huizache, abundant pirul, and maguey.

Modern Utilization. No cultural features are associated with this site, but in 1970 and 1972 it was noted that segments of this site were being utilized as garbage dumps for the municipio of San Juan Teotihuacan and the modern villages of San Juan Evangelista and Santa Maria Maquixco. The site area is used primarily for grazing purposes although a maguey plantation with associated bancals is found in the central and northern site area. Three small jagueys are found within the site.

Archaeological Remains. The total multicomponent site occupies 36.0 ha, while the Classic component occupies approximately 10.5 ha. Seventy-three mounds were identified of which 53 have some Classic occupation. The site has five phases of Preclassic, six phases of Classic, and four phases of Post Classic represented for a total of 15. Associated non-Classic sites include TF-138, TT-133, TA-28, and TA-219/221, while Classic site TC-7 is located immediately to the west, and Classic sites TC-11, TC-12, and TC-121 are located immediately to the east. The following mounds were identified:

TC-8: Pyramid, 380 m^2, slightly eroded; an excavated site. No marine or riverine mollusk shells were associated.

TC-8:1-2, 2,730 m^2, slightly eroded; a partially excavated site. Artifactual remains indicated all components represented. A total of 215 marine shells were found during excavation. (See Appendices I, II and III.)

TC-8:3, 3,650 m^2 slightly eroded; a partially excavated site. Artifactual remains indicated all components represented. A total of 3,927 marine shells were found during excavation. (See Appendices I, II and III.)

TC-8:4, 2,595 m^2, slightly eroded; a totally excavated site. Artifactual remains indicated that all components were represented. Twenty-nine marine shells were found during excavation. (See Appendix I.)

TC-8:5, 1,335 m^2, slightly eroded. Artifactual remains indicated traces of Early Tzacualli, traces of Late Tlamimilolpa, sparse to moderate Early and Late Xolalpan, sparse Mazapan, and sparse Aztec components.

TC-8:6, 1,452 m^2, slightly eroded. Artifactual remains indicated traces of Patlachique, traces of Early Tzacualli, traces of Early Tlamimilolpa, traces of Late Tlamimilolpa, moderate Early Xolalpan, sparse Late Xolalpan, traces of Metepec, traces of Mazapan, and sparse to moderate Aztec components.

TC-8:7, 1,025 m^2, slightly eroded. Artifactual remains indicated sparse Patlachique, sparse Early Tzacualli, sparse Miccaotli, sparse Early Tlamimilolpa, sparse Late Tlamimilolpa, moderate Early Xolalpan, sparse Late Xolalpan, traces of Mazapan, and moderate Aztec components.

TC-8:8, 1,843 m^2, slightly eorded. Artifactual remains indicated traces of Patlachique, traces of Early Tzacualli, sparse Late Tlamimilolpa, moderate Early Xolalpan, sparse Late Xolalpan, sparse Mazapan, and moderate Aztec components.

TC-8:9, 1,670 m^2, slightly eroded. Artifactual remains indicated traces of Patlachique, traces of Early Tzacualli, traces of Early Tlamimilolpa, sparse to moderate Late Tlamimilolpa, sparse Early Xolalpan, sparse Late Xolalpan, sparse Mazapan, and sparse to moderate Aztec components.

TC-8:10, 765 m^2, slightly eroded. Artifactual remains indicated traces of Early Tlamimilolpa, traces of Late Tlamimilolpa, sparse Early Xolalpan, and moderate Aztec components.

TC-8:11, 1,910 m^2, slightly eroded. Artifactual remains indicated traces of Early Tzacualli, sparse Late Tlamimilolpa, sparse Early Xolalpan, sparse Late Xolalpan, traces of Mazapan, and moderate to heavy Aztec components.

TC-8:12, 830 m^2, slightly to moderately eroded. Artifactual remains indicated traces of Early Tlamimilolpa, traces of Late Tlamimilolpa, sparse Early Xolalpan, sparse Late Xolalpan, sparse Mazapan, and moderate to heavy Aztec components.

TC-8:13, 3,285 m^2, slightly eroded. Artifactual remains indicated traces of Early Tzacualli, traces of Early Tlamimilolpa, traces of Late Tlamimilolpa, sparse Early Xolalpan, traces of Late Xolalpan, a possible trace of Late Mazapan, and extremely heavy Aztec components.

TC-8:14, 1,952 m^2, slightly to moderately eroded. Artifactual remains indicated traces of Patlachique, sparse Early Tzacualli, a trace of Early Tlamimilolpa, sparse Late Tlamimilolpa, sparse to moderate Early Xolalpan, sparse Late Xolalpan, a possible trace of Mazapan, and moderate Aztec components. One unidentified, worked fragment of marine shell was found during surface survey.

TC-8:15, 420 m^2, slightly to moderately eroded. Artifactual remains indicated traces of Early Tzacualli, traces of Early Tlamimilolpa, traces of Late Tlamimilolpa, traces of Early Xolalpan, a possible trace of Mazapan, heavy Aztec components, and modern debris.

TC-8:16, 327 m^2, slightly to moderately eroded. Artifactual remains indicated traces of Patlachique, sparse Early Tzacualli, traces of Early Xolalpan, traces of Mazapan, and heavy Aztec components.

TC-8:17, 525 m^2, slightly eroded. Artifactual remains indicated sparse Early Tzacualli, sparse Early Xolalpan, traces of Late Xolalpan, a trace of Mazapan, and heavy Aztec components.

TC-8:18, 547 m², slightly eroded. Artifactual remains indicated traces of Early Tzacualli, sparse Early Xolalpan, sparse Late Xolalpan, traces of Mazapan, and heavy Aztec components.

TC-8:19, 1,170 m², heavily eroded and damaged. Artifactual remains indicated traces of Early Xolalpan, sparse Mazapan, and heavy Aztec components.

TC-8:20, 405 m², slightly eroded. Artifactual remains indicated sparse Early Tzacualli, traces of Early Xolalpan, traces of Late Xolalpan, and heavy Aztec components.

TC-8:21, 592 m², slightly eroded. Artifactual remains indicated slight Late Tlamimilolpa, moderate Early Xolalpan, moderate Late Xolalpan, and moderate to heavy Aztec components.

TC-8:22, 864 m², slightly eroded Artifactual remains indicated traces of Early Tlamimilolpa, traces of Late Tlamimilolpa, sparse to moderate Early Xolalpan, sparse Late Xolalpan, sparse Mazapan, and sparse Aztec components.

TC-8:23, 660 m², slightly eroded. Artifactual remains indicated sparse Patlachique, sparse to moderate Early Tzacualli, a trace of late Tlamimilolpa, moderate Early Xolalpan, traces of Late Xolalpan, traces of Metepec, moderate Mazapan, and sparse to moderate Aztec components.

TC-8:24, 1,050 m², slightly eroded. Artifactual remains idicated traces of Patlachique, moderate Early Tzacualli, moderate Late Tlamimilolpa, sparse Early Xolalpan, moderate Late Xolalpan, traces of Metepec, sparse to moderate Mazapan, and moderate Aztec components.

TC-8:25, 1,190 m², slightly to moderately eroded. Artifactual remains indicated traces of Patlachique, sparse Early Tzacualli, sparse to moderate Early Xolalpan, traces of Late Xolalpan, sparse to moderate Mazapan, and heavy Aztec components.

TC-8:26, 408 m², slightly eroded Artifactual remains indicated traces of Early Tzacualli, traces of Late Xolalpan, sparse Mazapan, and extremely heavy Aztec components.

TC-8:27, 1,725 m², slightly eroded. Artifactul remains indicated sparse to moderate Early Xolalpan, traces of Late Xolalpan, and heavy Aztec components.

TC-8:28, 980 m², slightly eroded. Artifactual remains indicated traces of Patlachique, sparse Early Tzacualli, traces of Early Tlamimilolpa, sparse Late Tlamimilolpa, sparse to moderate Early Xolalpan, traces of Late Xolalpan, moderate to heavy Mazapan, and moderate Aztec components. One unworked fragment of a "Bittersweet Clam" was found during surface reconnaissance.

TC-8:29, 305 m², slightly eroded. Artifactual remains indicated traces of Early Tzacualli, sparse to moderate Mazapan, and extremely heavy Aztec components.

TC-8:30, 585 m², slightly eroded. Artifactual remains indicated traces of Patlachique, sparse to moderate Early Tzacualli, sparse Early Xolalpan, traces of Late Xolalpan, sparse to moderate Mazapan, and moderate Aztec components.

TC-8:31, 1,145 m², slightly eroded. Artifactual remains indicated traces of Patlachique, moderate to heavy Early Tzacualli, sparse Mazapan, and sparse to moderate Aztec components.

TC-8:32, 757 m², slightly eroded. Artifactual remains indicated sparse to moderate Mazapan and moderate to heavy Aztec components.

TC-8:33, 180 m², slightly eroded Artifactual remains indicated sparse to moderate Early Tzacualli, traces of Mazapan, and especially heavy Aztec components.

TC-8:34, 460 m², slightly eroded. Artifactual remains indicated traces of Early Tlamimilolpa, traces of Late Tlamimilolpa, sparse to moderate Mazapan, and moderate to heavy Aztec components.

TC-8:35, 445 m², slightly eroded. Artifactual remains indicated traces of Patlachique, sparse Early Tzacualli, traces of Late Tlamimilolpa, traces of Early Xolalpan, sparse Mazapan, and especially heavy Aztec components.

TC-8:36, 340 m², slightly eroded. Artifactual remains indicated traces of Early Tzacualli and especially heavy Aztec components.

TC-8:37, 830 m², slightly eroded. Artifactual remains indicated traces of Early Tzacualli, especially heavy Aztec components, and modern debris.

TC-8:38, 1,155 m², slightly eroded. Artifactual remains indicated traces of Patlachique, traces of

Early Tzacualli, traces of Miccaotli, traces of Early Tlamimilolpa, sparse Mazapan, and especially heavy Aztec components.

TC-8:39, 270 m^2, heavily damaged and eroded. Artifactual remains indicated sparse to moderate Early Tzacualli, traces of Early Xolalpan, traces of Late Xolalpan, traces of Mazapan, and heavy Aztec components.

TC-8:40, 147 m^2, heavily damaged and eroded. Atifactual remains indicated sparse to moderate Early Tzacualli, sparse Mazapan, and moderate Aztec components.

TC-8:41, 870 m^2, heavily damaged and eroded. Artifactual remains indicated a possible Early Tzacualli trace, a possible Mazapan trace, and especially heavy Aztec components.

TC-8:42, 315 m^2, heavily damaged and eroded. Artifactual remains indicated traces of Early Tzacualli, a possible Early Xolalpan trace, sparse Mazapan, and especially heavy Aztec components.

TC-8:43, 327 m^2, heavily damaged and eroded. Artifactual remains indicated traces of Mazapan and especially heavy Aztec components.

TC-8:44, 572 m^2, heavily damaged and eroded. Artifactual remains indicated traces of Late Tzacualli, a possible trace of Early Xolalpan, sparse Mazapan, and especially heavy Aztec components.

TC-8:45, 247 m^2, slightly eroded. Artifactual remains indicated sparse to moderate Early Tzacualli, sparse Early Xolalpan, sparse Mazapan, and moderate to heavy Aztec components.

TC-8:46, 105 m^2, slightly eroded. Artifactual remains indicated traces of Patlachique, sparse to moderate Early Tzacualli, sparse Miccaotli, sparse Late Tlamimilolpa, sparse Early Xolalpan, sparse Late Xolalpan, sparse to moderate Mazapan, and moderate Aztec components.

TC-8:47, 115 m^2, slightly to moderately eroded. Artifactual remains indicated sparse Patlachique, moderate Early Tzacualli, sparse Early Xolalpan, sparse to moderate Mazapan, and moderate to heavy Aztec components.

TC-8:48, 642 m^2, slghtly eroded. Artifactual remains indicated sparse Patlachique, sparse to moderate

Early Tzacualli, and moderate to heavy Aztec components.

TC-8:49, 245 m^2, heavily damaged and eroded. Artifactual remains indicated traces of Cuanalan, sparse Patlachique, sparse to moderate Early Tzacualli, traces of Early Tlamimilolpa, sparse Late Tlamimilolpa, sparse Mazapan, and sparse to moderate Aztec components.

TC-8:50, 685 m^2, slightly to moderately eroded. Artifactual remains indicated sparse to moderate Early Tzacualli and especially heavy Aztec components.

TC-8:51, 665 m^2, slightly eroded. Artifactual remains indicated traces of Cuanalan, sparse Patlachique, moderate to heavy Early Tzacualli and Late Tzacualli, traces of Early Tlamimilolpa, traces of Late Tlamimilolpa, sparse Early Xolalpan, traces of Mazapan, and sparse to heavy Aztec components, and modern debris.

TC-8:52, 510 m^2, slightly eroded. Artifactual remains indicated traces of Patlachique, sparse Early Tzacualli, traces of Early Xolalpan, traces of Late Xolalpan, sparse Mazapan, especially heavy Aztec components, and traces of modern debris.

TC-8:53, 670 m^2, slightly to moderately eroded. Artifactual remains indicated sparse Patlachique, sparse to moderate Early Tzacualli, sparse Mazapan, and sparse to moderate Aztec components.

TC-8:54, 287 m^2, slightly eroded. Artifactual remains indicated traces of Patlachique, sparse to moderate Early Tzacualli and Late Tzacualli, sparse Early Xolalpan, sparse to moderate Mazapan, and moderate to heavy Aztec components.

TC-8:55, 630 m^2, heavily eroded. Artifactual remains indicated sparse to moderate Early Tzacualli, traces of Late Tlamimilolpa, sparse Early Xolalpan, sparse to moderate Late Xolalpan, sparse Mazapan, and sparse Aztec components.

TC-8:56, 655 m^2, heavily eroded and damaged. Artifactual remains indicated sparse to moderate Late Tlamimilolpa, traces of Early Xolalpan, sparse to moderate Late Xolalpan, sparse to moderate Mazapan, and sparse to moderate Aztec components.

TC-8:57, 970 m^2, slightly to moderately eroded. Artifactual remains indicated traces of Patlachique, sparse Early Tzacualli, sparse Late Xolalpan, sparse Mazapan, and especially heavy Aztec components, and traces of modern debris.

TC-8:58, 505 m^2, slightly eroded. Artifactual remains indicated traces of Patlachique, sparse Early Tzacualli, traces of late Tlamimilolpa, traces of Early Xolalpan, traces of Late Xolalpan, sparse to moderate Mazapan, and moderate to heavy Aztec components.

TC-8:59, 587 m^2, slightly eroded. Artifactual remains indicated sparse Early Tzacualli, sparse to moderate Late Xolalpan, sparse Mazapan, especially heavy Aztec components, and sparse modern debris.

TC-8:60, 325 m^2, slghtly eroded. Artifactual remains indicated sparse Early Tzacualli, sparse to moderate Mazapan, moderate to heavy Aztec components, and sparse modern debris.

TC-8:61, 415 m^2, slightly eroded and damaged. Artifactual remains indicated sparse Early Tzacualli, sparse Early Xolalpan, sparse Late Xolalpan, moderate to heavy Mazapan, and heavy Aztec components.

TC-8:62, 470 m^2, slightly eroded. Artifactual remains indicated traces of Patlachique, sparse Early Tzacualli, sparse Late Tlamimilolpa, sparse to moderate Early Xolalpan, sparse to moderate Late Xolalpan, moderate to heavy Mazapan, and heavy Aztec components.

TC-8:63, 285 m^2, slightly eroded. Artifactual remains indicated traces of Patlachique, sparse to moderate Early and Late Tzacualli, traces of Early Tlamimilolpa, traces of Late Tlamimilolpa, sparse Early Xolalpan, and moderate to heavy Aztec components.

TC-8:64, 160 m^2, moderately eroded. Artifactual remains indicated traces of Early Tlamimilolpa, traces of Late Tlamimilolpa, traces of Early Xolalpan, sparse to moderate Mazapan, and especially heavy Aztec components.

TC-8:65, 172 m^2, moderately eroded. Artifactual remains indicated traces of Early Tlamimilolpa, traces of Late Tlamimilolpa, traces of Early Xolalpan, sparse to moderate Mazapan, and heavy Aztec components.

TC-8:66, 178 m^2, moderately eroded. Artifactual remains indicated traces of Early Tlamimilolpa,

traces of Late Tlamimilolpa, traces of Early
Xolalpan, sparse to moderate Mazapan, and heavy Aztec
components.

TC-8:67, 190 m^2, moderately eroded. Artifactual
remains indicated traces of Early Tlamimilolpa,
traces of late Tlamimilolpa, traces of Early
Xolalpan, sparse to moderate Mazapan, and moderate to
heavy Aztec components.

TC-8:68, 225 m^2, slightly to moderately eroded.
Artifactual remains indicated sparse to moderate
Early Tzacualli, traces of Late Xolalpan, sparse to
moderate Mazapan, and heavy Aztec components.

TC-8:69, 735 m^2, slightly eroded. Artifactual
remains indicated sparse to moderate Early Tzacualli,
traces of Late Tlamimilolpa, sparse to moderate
Mazapan, and a moderate to heavy Aztec components.

TC-8:70, 865 m^2, slightly eroded and damaged.
Artifactual remains indicated traces of Patlachique,
sparse to moderate Early Tzacualli, moderate to heavy
Mazapan, and moderate to heavy Aztec components.

TC-8:71, 890 m^2, heavily damaged and eroded.
Artifactual remains indicated sparse Patlachique,
sparse to moderate Early Tzacualli, sparse to
moderate Mazapan, and moderate to heavy Aztec
components.

TC-8:72, 505 m^2, slightly to moderate eroded.
Artifactual remains indicated heavy Early Tzacualli.

TC-8:73, 700 m^2, moderately to heavily eroded and
damaged. Artifactual remains indicated sparse to
moderate Early Tzacualli, possible traces of Late
Tlamimilolpa, sparse to moderate Mazapan, sparse to
moderate Aztec components, and sparse modern debris.

The site has a moderate to heavy distribution of lithic
material including obsidian blades, cores, projectile
points, and scrapers. Some chunks and cobbles of
obsidian were reported. Ground stone tools, especially
mano and metate fragments and mortar and pestle
fragments, were also noted. Other special lithic
materials included basalt, serpentine, and quartzite
artifacts.

Classification. TC-8 was interpreted by Marino
(1965:147, 164, 169) as a compound village with
quadrangular arrangements of houses dating to Middle and
Late Teotihuacan. The site also was one of his aligned
east-west sites of Middle to Late Teotihuacan associated

with TC-7, TC-25, TC-87, and TC-119. Marino (1975:303), in discussing the TT-133 component of this site, considered it to be a dispersed low-density Mazapan village. Sanders (1965:104, 107-116, 120-121; 1966) considered the TC-8 site to be an excellent example of a Classic village. The site is reclassified as a Large Nucleated Village with a probable population of between 300-600 according to Sanders during Xolalpan times. The site probably progressed from a Hamlet during the Late Preclassic, to a Small Nucleated Village during the Early Classic, to a Large Nucleated Village during the Late Terminal Classic. The site TC-8 has also been used by Sanders (1966:123, 148) as an example of a "typical" Classic village.

TC-10

Background. This site, also known as Venta de Carpio, was surveyed by Kolb in 1963 and resurveyed by Kolb in 1970 and 1972. The artifact collections from the Classic survey were processed by Kolb and Senulis in 1963-64 and consisted of five surface samples plus a large excavated sample. The site was initially reported by Tolstoy (1958) as a Preclassic site. Collections made by the TF and TT survey teams were also studied by Kolb. The ceramic materials from the Classic survey were excellent for chronological purposes, while figurine data were excellent and the Classic field reports were of excellent quality. Published and unpublished materials on this site include Tolstoy (1958) and Marino (1965, 1975).

Natural Setting. Site TC-10 is located in the Delta of the Teotihuacan Valley at 2,238 m and is in the Alluvial Plain Ecological Zone. Soils in the site area have a loamy to sandy texture with some silt and are medium to dark brown in color with traces of salt; depths range up to 185+ cm. There is moderate erosion in the northern site area, and some tepetate is exposed on a roadway in the northeastern section of the site. Light concentrations of rock and tezontle fragments are found in the area. Vegetationin the area includes grasses, nopal, and some maguey.

Modern Utilization. Cultural features in 1963 included three houses located to the south and southwest of the site. By 1970, the entire area had been damaged by the construction of several larger houses and a large granja. The site area in 1963 was used primarily for grazing purposes although a small garden plot was also to be found in the vicinity. A dirt road is located to the south of the site, and the Venta de Carpio-Teotihuacan Highway is located to the north of the site. Numerous ditches and small canals for drainage purposes had been excavated through most of the site area as of 1963.

Archaeological Remains. The total multicomponent site occupies 10.0 ha, while the Classic component occupies 5.2 ha. Over nine mounds were identified of which at least seven have Classic occupation. The site has five phases of Preclassic, five phases of Classic, and two phases of Post Classic represented for a total of 12. Associated non-Classic sites include TF-5, TT-4, TT-148, and TA-9, while Classic sites TC-122 and TC-3 are located to the west. The following mounds were identified:

TC-10:1, 1,330 m², moderately damaged in 1963; totally destroyed by 1970. Artifactual remains indicated sparse Miccaotli, sparse to moderate Early Tlamimilolpa, sparse Late Tlamimilolpa, sparse to moderate Mazapan, and heavy Aztec components.

TC-10:2, 385 m², moderately damaged n 1963; totally destroyed by 1972. Artifactual remains indicated traces of Miccaotli, moderate Lake Tlamimilolpa, traces of Mazapan, and heavy Aztec components. Three marine shells were recovered during excavation. (See Appendices I, II and III.)

TC-10:3, 107 m², totally destroyed by 1970. Artifactual remains indicated traces of Miccaotli.

TC-10:4, 115 m², totally destroyed by 1970. Artifactual remains indicated traces of Miccaotli.

TC-10:5, 365 m², moderately destroyed in 1963; totally destroyed by 1970. Artifactual remains indicated sparse to moderate Cuanalan, sparse Miccaotli, sparse Late Tlamimilolpa, sparse Early Xolalpan, traces of Mazapan, and moderate Aztec components.

TC-10:6, 185 m², heavily damaged in 1963; totally destroyed by 1970. Artifactual remains indicated sparse Miccaotli, sparse Early Tlamimilolpa, sparse Late Tlamimilolpa, and moderate Aztec components.

TC-10:7, 138 m², heavily damaged in 1963; totally destroyed by 1970. Artifactual remains indicated sparse Early Tlamimilolpa, sparse Late Tlamimilolpa, traces of Mazapan, and sparse to moderate Aztec components.

The site has a sparse distribution of lithic materials including obsidian blades and scrapers plus some chunks and flakes of obsidian. No other lithic materials have been recorded in the surface survey data.

Classification. TC-10 was interpreted by Marino (1965:170, 1975:152) as an isolated non-maize farming workshop site, probably a salt-making locality, during Middle and Late Classic. This site is reclassified as a Small Nucleated Village with a probable population of between 100-200 individuals during Xolalpan times. The site probably also served as a salt-making site during Middle Classic. This site probably progressed from a Hamlet or Specialized Activity Locus during the Early Classic to a Small Nucleated Village with associated specialized activities during the Middle Classic and was apparently abandoned in Late Terminal Classic.

TC-13

Background. This site, also known as San Lorenzo Tlamimilolpa, was surveyed by Kolb in 1963 and resurveyed by Kolb in 1970 and 1972. The artifact collections from the Classic survey were processed by Kolb in 1963 and 1964 and consisted of seven samples. Collections made by TT and TA survey teams were also studied by Kolb. The ceramic materials from the Classic survey were of excellent quality for chronological purposes, while figurine data were good and the Classic field reports were of excellent quality. Published and unpublished materials on this site include Marino (1965, 1975), Sanders (1965) and Millon (1973).

Natural Setting. Site TC-13 is located in the Middle Valley, South Piedmont, of the Teotihuacan Valley between 2,270-2,280 m in the Lower Piedmont Ecological Zone. Soils in the site area have a sandy to loamy texture and are tan to light brown in color, with depths ranging from 0 to 120+ cm. There is moderate erosion and alteration in the western area of the site, and tepetate was exposed in the western section of the site in 1963. Moderate to heavy concentrations of rock and tezontle fragments are found in the area. Vegetation in the vicinity includes various grasses and some nopal. Other natural features include a wash-road to the south of the site.

Modern Utilization. Cultural features include no structures. The site area is used for agricultural purposes, especially including cultivation of maize in 1963. The site has since been particularly altered but is still used for maize as of 1972. A federal highway from Xometla to San Lorenzo transects the western side of the site. Since 1963, the western segment of the site has been seriously damaged by the construction of the toll road (autopista) to San Juan Teotihuacan and the archaeological zone. An electric power line running east to west is found immediately to the south of the site, as are several roads and paths.

Archaeological Remains. The total multicomponent site occupies 13.6 ha, while the Classic component occupies the same area. Four mounds were identified of which three have Classic occupation. The site has five phases of Preclassic, six phases of Classic, and three phases of Post Classic represented for a total of 14. Associated non-Classic sites include TF-140, TT-23, TT-161, TA-23, TA-169-17, and TA-170-40. Associated Classic sites include TC-14, and TC-117 is located to the east, TC-9 to the southwest, and TC-12 and 121 to the north. The following mounds were identified:

TC-13:1, 6,240 m², moderately eroded. Artifactual remains indicated moderate Early Tzacualli, traces of Miccaotli, moderate Early Tlamimilolpa, sparse Late Tlamimilolpa, sparse to moderate Xolalpan, sparse to moderate Mazapan, and sparse Aztec components.

TC-13:2, 1,755 m², partally damaged by 1963 and almost totally destroyed by 1970. Artifactual remains indicated traces of Cuanalan, traces of Tezoyuca, moderate Early Tzacualli, sparse Late Tzacualli, sparse Miccaotli, sparse Early Tlamimilolpa, sparse to moderate Late Tlamimilolpa, sparse to moderate Early and Late Xolalpan, traces of Metepec, sparse to moderate Xometla, sparse to moderate Mazapan, and sparse to moderate Aztec components.

TC-13:3, 4,185 m², moderately eroded in 1963; seriously damaged by 1970. Artifactual remains indicated sparse to moderate Early and Late Tzacualli, traces of Miccaotli, sparse to moderate Early Tlamimilolpa, sparse to moderate Late Tlamimilolpa, sparse to moderate Early Xolalpan, sparse to moderate Xometla, sparse to moderate Mazapan, and sparse to moderate Aztec components. One unworked specimen of a *Chama echinata* was found during surface survey. (See Appendices I, II and III.)

The site has a sparse distribution of lithic materials including obsidian projectile points and scrapers.

Site TC-13 was included by Millon (1973) within the urban metropolis. The following correspondences can be recognized between the results of the Teotihuacan Mapping Project (TMP) and the Teotihuacan Valley Project (TVP). Millon's maps provided site and coordinate data (1973:140), while site dimensions were calculated by Kolb from these maps:

Site Numbers		Area TMP Maps (m²)	Area TVP Maps (m²)
TC-13:1	A = S6W4:1F	400	6,240
TC-13:2	B = S6W4:1D	1,764	1,755
TC-13:3	C = S6W4:1E	4,400	4,185
		6,564 m²	12,180 m²

Difference: 5,616 m²

The total site area for TC-13 was 13.6 ha, which does not correspond closely with Millon's 15.5 ha. It appears that the area differences for individual mounds is a function of the serious alternation of TC-13:1 and 2 when the autopista was constructed through the site in 1964-1965. Millon's survey team did not investigate this section until after the highway construction had been completed.

Classification. TC-13 was interpreted by Marino (1965:172) as a proto-Toltec living site with associated canal irrigation systems. Marino (1975:217-219, 294-295), in considering the TT-23 component of the site, indicated that it is a small nucleated village during the Xometla Phase and a small dispersed village during the Mazapan Phase. Sanders (1965:106) also considered the site to be associated with the Toltec Period. This site is reclassified as a Barrio associated with a Supra-Regional Center, and a population is, therefore, not calculated. The site probably progressed from a Small Nucleated Village during the Early Classic to a Barrio during the Middle Classic, and either a Barrio or Small Nucleated Village during the Late/Terminal Classic.

TC-40

Background. This site, also known as Tlacatele de San Juan Teacalco, was surveyed by Marino in 1963 and has not been resurveyed. The artifact collections from the Classic survey were processed by Kolb in 1963 and 1964 and consisted of 23 samples. Collections made by TF and TA survey teams were also studied by Kolb. The ceramic materials from the Classic survey were excellent for chronological purposes, while figurine data were excellent and the Classic field reports were of good quality. Published and unpublished materials on this site include Marino (1965) and Sanders (1965).

Natural Setting. Site TC-40 is located in the Cerro Gordo North Slope of the Teotihuacan Valley between 2,400-2,340 m in the Middle Piedmont Ecological Zone. Soils in the site area have a loamy to sandy texture and are medium to dark brown in color, with depths ranging up to 110 cm. There is slight erosion in the north edge of the site where tepetate is exposed. Moderate to heavy concentrations of rock and tezontle fragments are found in the area. Vegetation in the vicinity includes pirul, maguey, huizache, nopal, and various grasses. Other natural features include a wash along the north and northeastern edges of the site; an incipient wash is located at the western margin.

Modern Utilization. Cultural features include the Panteon Teacalco at the Northeast corner of the site. The site area is used primarily for agricultural purposes including the cultivation of maize and barley. The entire area is composed of a series of stone terraces and maguey bancals. Several drainage ditches along the western boundary of the site were noted.

Archaeological Remains. The total multicomponent site occupies 16.5 ha, while the Classic component occupies 9.5 ha. Twenty-two mounds were identified of which nine have Classic occupation. The site has three phases of the Preclassic, six phases of the Classic, and two phases of the Post Classic represented for a total of 11. Associated non-Classic sites include TF-16 and TA-44, while Classic sites TC-41, 42, and 47 are located to the southeast and TC-20 to the northwest. The following mounds were identified:

> TC-40:1, 215 m^2, slightly eroded. Artifactual remains indicated sparse Early Tzacualli, moderate to heavy Early Tlamimilolpa, sparse to moderate Late Tlamimilolpa, sparse Early Xolalpan, traces of Late Xolalpan, and traces of Aztec components.

TC-40:2, 515 m^2, slightly eroded. Artifactual remains indicated possible traces of Cuanalan, sparse Early Tzacualli, traces of Miccaotli, sparse Early Tlamimilolpa, moderate Late Tlamimilolpa, moderate Early Xolalpan, sparse to moderate Late Xolalpan, and traces of Aztec components.

TC-40:3, 330 m^2, slightly eroded. Artifactual remains indicated traces of Late Tzacualli, moderate Early Tlamimilolpa, moderate Late Tlamimilolpa, sparse to moderate Early Xolalpan, and sparse Aztec components.

TC-40:4, 345 m^2, slightly eroded. Artifactual remains indicated sparse to moderate Early Tzacualli, traces of Miccaotli, moderate Late Tlamimilolpa, moderate Early Xolalpan, sparse Late Xolalpan, traces of Mazapan, and moderate Aztec components.

TC-40:5, 235 m^2, heavily eroded and damaged. Artifactual remains indicated sparse Early Tzacualli, sparse Miccaotli, heavy Early Tlamimilolpa, moderate to heavy Late Tlamimilolpa, sparse Early Xolalpan, sparse Late Xolalpan, traces of Metepec, traces of Mazapan, and moderate Aztec components.

TC-40:6, 295 m^2, slightly eroded. Artifactual remains indicated possible traces of Cuanalan, traces of Early Tlamimilolpa, traces of Late Tlamimilolpa, sparse Early Xolalpan, a trace of Mazapan, and heavy Aztec components.

TC-40:7, 180 m^2, slightly eroded. Artifactual remains indicated traces of Early Tzacualli, traces of Miccaotli, sparse Early Tlamimilolpa, sparse to moderate Late Tlamimilolpa, sparse to moderate Early Xolalpan, sparse Late Xolalpan, sparse to moderate Mazapan, and moderate to heavy Aztec components. One unidentified, unworked shell fragment was recovered during surface reconnaissance.

TC-40:8, 330 m^2, heavily eroded. Artifactual remains indicated traces of Early Tzacualli, traces of Miccaotli, heavy Early Tlamimilolpa, heavy Late Tlamimilolpa, moderate Early Xolalpan, traces of Late Xolalpan, and sparse Aztec components.

TC-40:9, 300 m^2, heavily eroded and damaged. Artifactual remains indicated possible traces of Cuanalan, moderate Early Tlamimilolpa, heavy Late Tlamimilolpa, moderate Early Xolalpan, and moderate Aztec components.

TC-40:10, size unrecorded, slightly eroded. Artifactual remains indicated traces of Early Tzacualli, moderate Early Tlamimilolpa, moderate to heavy Late Tlamimilolpa, sparse to moderate Early Xolalpan, traces of Mazapan, and moderate Aztec components.

TC-40:11, size unrecorded, slightly eroded. Artifactual remains indicated moderate Early Tlamimilopa, moderate Late Tlamimilolpa, sparse Early Xolalpan, and sparse to moderate Aztec components.

TC-40:12, size unrecorded, slightly eroded. Artifactual remains indicated sparse to moderate Early Tzacualli, sparse Early Tlamimilolpa, sparse Late Tlamimilolpa, moderate Early Xolalpan, moderate Late Xolalpan, and sparse to moderate Aztec components.

TC-40:13, size unrecorded slightly eroded. Artifactual remains indicated sparse Early Tlamimilolpa, moderate Late Tlamimilolpa, sparse Early Xolalpan, and sparse Aztec components.

TC-40:14, size unrecorded, moderately eroded. Artifactual remains indicated sparse Early Tlamimilolpa, moderate Late Tlamimilolpa, sparse Early Xolalpan, traces of Late Xolalpan, and sparse to moderate Aztec components.

TC-40:15, size unrecorded, moderately eroded and damaged. Artifactual remains indicated sparse Late Tzacualli, sparse Miccaotli, sparse to moderate Early Tlamimilolpa, sparse to moderate Late Tlamimilolpa, sparse Early Xolalpan, possible traces of Mazapan, and sparse to moderate Aztec components.

TC-40:16, size unrecorded, slightly eroded. Artifactual remains indicated sparse Early Tzacualli, sparse Miccaotli, moderate to heavy Early Tlamimilolpa, moderate to heavy Late Tlamimilolpa, traces of Early Xolalpan, and sparse Aztec components.

TC-40:17, size unrecorded, slightly to moderately eroded. Artifactual remains indicated traces of Early Tzacualli, possible traces of Miccaotli, moderate Early Tlamimilolpa, moderate Late Tlamimilolpa, sparse to moderate Early Xolalpan, sparse Late Xolalpan, and sparse to moderate Aztec components.

TC-40:18, size unrecorded, moderately eroded. Artifactal remains indicated sparse Early Tzacualli, traces of Miccaotli, moderate Early Tlamimilolpa, moderate Late Tlamimilolpa, sparse Early Xolalpan, traces of Late Xolalpan, and moderate Aztec components.

TC-40:19, size unrecorded, slightly eroded. Artifactual remains indicated moderate Early Tzacualli, traces of Miccaotli, moderate Early Xolalpan, traces of Late Xolalpan, and sparse to moderate Aztec components.

TC-40:20, size unrecorded, slight erosion assumed. Artifactual remains indicated sparse Late Tzacualli, traces of Miccaotli, moderate Early Tlamimilolpa, moderate Late Tlamimilolpa, sparse Early Xolalpan, traces of Late Xolalpan, and traces of Mazapan components.

TC-40:21, size unrecorded, moderate erosion assumed. Artifactual remains indicated moderate to heavy Late Tzacualli, sparse to moderate Early Tlamimilolpa, sparse to moderate Late Tlamimilolpa, and traces of Aztec components.

TC-40:22, size unrecorded, slight erosion assumed. Artifactual remains indicated traces of Early Tzacualli, traces of Miccaotli, sparse Early Tlamimilolpa, sparse Late Tlamimilolpa, and traces of Early Xolalpan components.

The site has an abundant distribution of lithic materials including obsidian blades, cores, and scrapers. Ground stone tools, especially metates, mortars, and pestles, were also noted. Two stone rings and a fragment of a Huehueteotl stone statue were also found.

Classification. TC-40 was interpreted by Marino (1965:141, 145-147) as a Tlamimilolpa and Xolalpan apartment cluster. Sanders (1965:107, 116, 120-121, 176) has regarded the site as a small town on the Cerro Gordo North Slope. This site is reclassified as a Provincial Center with a probable population between 750-1,500 during Xolalpan times. The site probably was either a Large Nucleated Village or a Provincial Center (or Regional Center ?) during the Early Classic and became a Provincial Center during the Middle Classic, and probably reverted to a lesser status during the Late and Terminal Classic in that Late Classic artifacts are not abundant at the site.

TC-49

Background. This site, also known as Tenango de Santa Maria Maquisco el Alto, was surveyed by Marino in 1963 and was not resurveyed. The artifact collections from the Classic survey were processed by Kolb in 1964 and 1965 and consisted of 28 samples. Collections made by the TF survey team were also studied by Kolb. The ceramic materials from the Classic survey were excellent for chronological purposes, while figurine data were poor and the Classic field reports of poor quality. Published and unpublished materials on this site include Marino (1965) and Sanders (1965).

Natural Setting. Site TC-49 is located in the Cerro Gordo North Slope area of the Teotihuacan Valley between 2,450-2,470 m in the Middle Piedmont Ecological Zone. Soils in the site area have a sandy to loamy texture and are tan to medium brown in color with depths ranging from 0-35 cm. There is moderate erosion in most of the site area, and tepetate is exposed along the northern edge of the site. Moderate concentrations of rock and tezontle fragments are found in the area. Vegetation in the vicinity includes pirul, nopal, maguey, huizache, and various grasses. Other natural features include several incipient washes and barrancas to the east and west of the site.

Modern Utilization. Cultural features include no known structures or jagueys. The site area is used primarily for agricultural purposes including the cultivation of maguey and maize. Stone terraces and maguey bancals are also found in the vicinity.

Archaeological Remains. The total multicomponent site occupies 9.2 ha, while the Classic component occupies 9.2 ha. Twenty-nine mounds were identified of which all have Classic occupation. The site has three phases of the Preclassic, five phases of the Classic, and two phases of the Post Classic represented for a total of ten. Associated non-Classic sites include TF-144, while Classic sites TC-40, TC-41, and TC-47 are located to the northwest and TC-58 to the southeast. It is unfortunate that no mound sizes or erosional data were recorded during this survey. Data, however, are available as to the phase occupations of the 29 mounds.

TC-49:1. Artifactual remains indicated sparse Ealy Tzacualli, sparse Miccaotli, moderate Early Tlamimilolpa, moderate Late Tlamimilolpa, possible traces of Early Xolalpan, traces of Mazapan, and traces of Aztec components. One unidentified shell fragment was recovered during test excavations. (See Appendix I.)

TC-49:2. Artifactual reamins indicated sparse Early Tzacualli, sparse Late Tzacualli, sparse to moderate Miccaotli, moderate to heavy Early Tlamimilolpa, moderate Late Tlamimilolpa, traces of Early Xolalpan, sparse Mazapan, and sparse to moderate Aztec components.

TC-49:3. Artifactual remains indicated possible traces of Cuanalan, sparse Miccaotli, moderate Early Tlamimilolpa, moderate Late Tlamimilolpa, sparse Mazapan, and sparse Aztec components. One worked Unionidae (Unio), "Freshwater Clam" fragment was found during testing operations. (See Appendices I, II and III.)

TC-49:4. Artifactual remans indicated traces of Early Tzacualli, traces of Late Tzacualli, sparse Miccaotli, moderate Early Tlamimilolpa, moderate Late Tlamimilolpa, traces of Mazapan, and sparse to moderate Aztec components.

TC-49:5. Artifactual remains indicated traces of Early Tzacualli, moderate Late Tzacualli, sparse Miccaotli, moderate Early Tlamimilolpa, moderate Late Tlamimilolpa, traces of Mazapan, and sparse to moderate Aztec components.

TC-49:6. Artifactual remains indicated possible traces of Early Tzacualli, traces of Late Tzacualli, sparse Miccaotli, sparse to moderate Early Tlamimilolpa, moderate Late Tlamimilolpa, traces of Early Xolalpan, sparse to moderate Mazapan, and traces of Aztec components.

TC-49:7. Artifactual remains indcated traces of Early Tzacualli, sparse Late Tzacualli, sparse Miccaotli, moderate Early Tlamimilolpa, moderate Late Tlamimilolpa, sparse Mazapan, and traces of Aztec components.

TC-49:8. Artifactual remains indicated sparse Early Tzacualli, sparse Late Tzacualli, sparse Miccaotli, moderate Early Tlamimilolpa, moderate Late Tlamimilolpa, possible traces of Early Xolalpan, moderate Mazapan, and traces of Aztec components.

TC-49:9. Artifactual remains indicated sparse Miccaotli, moderate to heavy Early Tlamimilolpa, moderate Late Tlamimilolpa, possible traces of Early Xolalpan, possible traces of late Xolalpan, sparse Mazapan, and sparse Aztec components.

TC-49:10. Artifactual remains indicated sparse Early Tzacualli, sparse Late Tzacualli, traces of Miccaotli, moderate Early Tlamimilolpa, moderate Late Tlamimilolpa, possible traces of Early Xolalpan, and traces of Aztec components.

TC-49:11. No sample was taken, but observations indicated probable Tlamimilolpa occupations.

TC-49:12. Artifactual remains indicated traces of Early Tzacualli, traces of Late Tzacualli, sparse Miccaotli, moderate to heavy Early Tlamimilolpa, moderate to heavy Late Tlamimilolpa, sparse Mazapan, and sparse Aztec components.

TC-49:13. Artifactual remains indcated sparse to moderate Early Tzacualli, sparse Late Tzacualli, moderate Miccaotli, heavy Early Tlamimilolpa, sparse to moderate Late Tlamimilolpa, possible traces of Early Xolalpan, sparse Mazapan, and sparse Aztec components.

TC-49:14. Artifactual remains indicated traces of Early Tzacualli, traces of Late Tzacualli, moderate Miccaotli, moderate Early Tlamimilolpa, heavy Late Tlamimilolpa, possible traces of Early Xolalpan, moderate Mazapan, and sparse Aztec components.

TC-49:15. No sample was taken.

TC-49:16. Artifactual remains indicated sparse Early Tzacualli, sparse Late Tzacualli, sparse Miccaotli, moderate Early Tlamimilolpa, sparse to moderate Late Tlamimilolpa, moderate to heavy Mazapan, and moderate Aztec components.

TC-49:17. Artifactual remains indicated traces of Early Tzacualli, traces of Late Tzacualli, sparse Miccaotli, heavy Early Tlamimilolpa, heavy Late Tlamimilolpa, possible traces of Early Xolalpan, traces of Mazapan, and traces of Aztec components.

TC-49:18. Artifactual remains indicated traces of Early Tzacualli, sparse Late Tzacualli, sparse Miccaotli, moderate Early Tlamimilolpa, moderate Late Tlamimilolpa, possible traces of Early Xolalpan, sparse Mazapan, and sparse Aztec components.

TC-49:19. Artifactual remains indicated moderate Early Tlamimilolpa, sparse to moderate Late Tlamimilolpa, and sparse to moderate Mazapan components.

TC-49:20. Artifactual remains indcated traces of Miccaotli, moderate Early Tlamimilolpa, moderate Late Tlamimilolpa, sparse to moderate Early Xolalpan, and traces of Mazapan components.

TC-49:21. Artifactual remains indicated sparse Early Tlamimilolpa, sparse Late Tlamimilolpa, heavy Mazapan, and traces of Aztec components.

TC-49:22. Artifactual remains indicated sparse Early Tzacualli, sparse Late Tzacualli, traces of Miccaotli, sparse to moderate Early Tlamimilolpa, sparse to moderate Late Tlamimilolpa, sparse to moderate Mazapan, and sparse Aztec components.

TC-49:23. Artifactual remains indicated traces of Early Tlamimilolpa, heavy Mazapan, and sparse Aztec components.

TC-49:24. Artifactual remains indicated traces of Early Tzacualli, traces of Late Tzacualli, sparse Miccaotli, heavy Early Tlamimilolpa, heavy Late Tlamimilolpa, sparse Early Xolalpan, and traces of Mazapan components.

TC-49:25. Artifactual remains indicated sparse Early Tlamimilolpa, sparse to moderate Late Tlamimilolpa, possible traces of Early Xolalpan, heavy Mazapan, and sparse Aztec components.

TC-49:26. Artifactual remains indicated sparse Early Tlamimilolpa, sparse Late Tlamimilolpa, possible sparse Early Xolalpan, and traces of Mazapan components.

TC-49:27. Artifactual remains indicated traces of Miccaotli, heavy Early Tlamimilolpa, moderate to heavy Late Tlamimilolpa, moderate Early Xolalpan, sparse Mazapan, and sparse Aztec components.

TC-49:28. Artifactual remains indicated sparse Early Tlamimilolpa, sparse to moderate Late Tlamimilolpa, sparse to moderate Mazapan, and sparse Aztec components.

TC-49:29. Artifactual remains indicated moderate Early Tlamimilolpa, moderate Late Tlamimilolpa, possible traces of Early Xolalpan, traces of Mazapan, and sparse Aztec components.

The site has a moderate and discontinuous distribution of lithic materials including obsidian blades, cores, and scrapers. Ground stone tools, especially manos, metates, mortars, and pestles, were also noted.

Classification. TC-49 was interpreted by Marino (1965:164) as a compound village with a quadrangular arrangement of houses dating to Middle and Late Classic. Sanders (1965:104, 115-118, 120-121) considered the site to be a village. This site is reclassified as a Large Nucleated Village with a probable population between 600-1,200 during Xolalpan times. The site probably progressed from a Small Nucleated Village during the Early Classic to a Large Nucleated Village during the Middle Classic, and reverted either to a Hamlet or was abandoned during the Late/Terminal Classic.

TC-73

Background. This site, also known as Los Cuecillos de San Cristobal Colhuacan o Los Mogotes de Buena Vista, was surveyed by Marino in 1963 and has not been resurveyed. The artifacts from the Classic survey were processed by Kolb in 1964 and 1965 and consisted of 58 samples. Collections made by the TA survey team were also studied by Kolb. Ceramic materials from the Classic survey were excellent for chronological purposes, while figurine data were good and the Classic field reports were of good quality. Published and unpublished materials on this site include Marino (1965, 1975) and Sanders (1965).

Natural Setting. Site TC-73 is located in the Cerro Gordo North Slope area of the Teotihuacan Valley between 2,445-1,460 m in the Upper Piedmont Ecological Zone. Soils in the site have a loamy texture and are medium to dark brown in color with some gray soils also recorded. Soil depths range from 0-45+ cm. There is moderate erosion in the site area, and tepetate is exposed along the northern edge of the site. Moderate to heavy concentrations of rock and tezontle fragments are found in the area. Vegetation in the vicinity includes pirul, nopal, huizache, and various grasses. Marino indicates that the abundant vegetation at this site gives it the appearance of a park. Other natural features include a barranca to the north of the site and an incipient wash to the northeast and southeast.

Modern Utilization. Cultural features apparently include no structures and no jagueys. A possible abandoned silted-in jaguey is located northeast of the site. The area is used primarily for agricultural purposes including the cultivation of maize, beans, barley, and maguey; some grazing is also conducted in the vicinity. A canal or canalized barranca segment is located northwest of the site. Remnants of some stone terraces and maguey bancals are found to the south and northeast.

Archaeological Remains. The total multicomponent site occupies 21.0 ha, while the Classic component occupies 21.0 ha. One hundred and one mounds or concentrations of archaeological remains were identified all of which have Classic occupation. The site has four phases of the Preclassic, five phases of the Classic, and three phases of the Post Classic represented for a total of 12. Associated non-Classic sites include TF-145, TC-146, TT-134, and TA-124, while Classic sites TC-75, TC-78, and TC-79 are located to the north and Classic sites TC-76 and TC-77 are located to the south. The following mounds were identified:

TC-73:1, 840 m^2, slghtly eroded. Artifactual remains indicated traces of Early Tzacualli, sparse Early Tlamimilolpa, traces of Late Tlamimilolpa, traces of Mazapan, and moderate Aztec components.

TC-73:2, 870 m^2, slightly eroded. Artifactual remains indicated possible traces of Patlachique, sparse Early Tlamimilolpa, traces of Mazapan, and moderate Aztec components.

TC-73:3, 1,150 m^2, slightly eroded. No sample was taken.

TC-73:4, 430 m^2, slightly eroded. Artifactual remains indicated sparse Early Tlamimilolpa, moderate Late Tlamimilolpa, traces of Early Xolalpan, and sparse Aztec components.

TC-73:5, 1,520 m^2, slightly eroded. Artifactual remains indicated traces of Early Tlamimilolpa, moderate to heavy Aztec, and traces of modern components.

TC-73:6, 155 m^2, slightly eroded. Artifactual remains indicated traces of Early Tzacualli, traces of Early Tlamimilolpa, sparse to moderate Late Tlamimilolpa, sparse Early Xolalpan, traces of Late Xolalpan, sparse Mazapan, and sparse Aztec components.

TC-73:7, 75 m^2, slightly eroded. Artifactual remains indicated traces of Early Tzacualli, traces of Late Tzacualli, traces of Miccaotli, sparse Early Tlamimilolpa, sparse Late Tlamimilolpa, sparse Early Xolalpan, possible traces of Late Xolalpan, sparse Mazapan, and sparse Aztec components.

TC-73:8, 35 m^2, slightly eroded. Artifactual remains indicated sparse Early Tlamimilolpa, sparse Early Xolalpan, traces of Late Xolalpan, and sparse to moderate Aztec components.

TC-73:9, 65 m^2, slightly eroded. Artifactual remains indicated traces of Early Tlamimilolpa, sparse Late Tlamimilolpa, sparse Early Xolalpan, traces of late Xolalpan, and sparse Aztec components.

TC-73:10, 45 m^2, slightly eroded. Artifactual remains indicated traces of Early Tzacualli, traces of late Tzacualli, traces of Miccaotli, traces of Early Tlamimilolpa, sparse Late Tlamimilolpa, sparse

Early Xolalpan, possible traces of Late Xolalpan, and sparse Aztec components.

TC-73: 11, 85 m², slightly eroded. Artifactual remains indicated possible traces of Early Tzacualli, traces of Miccaotli, moderate Early Tlamimilolpa, heavy Late Tlamimilolpa, sparse Early Xolalpan, traces of late Xolalpan, and traces or Aztec components.

TC-73: 12, 350 m², slightly eroded. Artifactual remains indicated traces of Early Tzacualli, traces of Late Tzacualli, traces of Miccaotli, sparse Early Tlamimilolpa, moderate to heavy Late Tlamimilolpa, sparse Early Xolalpan, traces of Late Xolalpan, and traces of Aztec components.

TC-73: 13, 210 m², slightly eroded. Artifacual remains indicated sparse Early Tlamimilolpa, heavy Late Tlamimilolpa, heavy Early Xolalpan, sparse Late Xolalpan, and traces of Aztec components.

TC-73: 14, 910 m², slightly eroded. Artifactual remains indicated traces of Early Tzacualli, sparse Early Tlamimilolpa, heavy Late Tlamimilolpa, moderate to heavy Early Xolalpan, sparse Late Xolalpan, and sparse Aztec components.

TC-73: 15, 550 m², slightly eroded. Artifactual remains indicated traces of Early Tlamimilolpa, sparse Late Tlamimilolpa, moderate Early Xolalpan, moderate Late Xolalpan, and traces of Aztec components.

TC-73: 16, 645 m², slightly eroded. Artifactual remains indicated traces of Early Tzacualli, moderate Early Tlamimilolpa, moderate to heavy Late Tlamimilolpa, traces of Early Xolalpan, and sparse Aztec components.

TC-73: 17, 525 m², slightly eroded. Artifactual remains indicated sparse Early Tlamimilolpa, sparse to moderate Late Tlamimilolpa, sparse Early Xolalpan, traces of Late Xolalpan, and sparse Aztec components.

TC-73: 18, 4,050 m², heavily eroded and damaged. Artifactual remains indicated sparse Early Tlamimilolpa, sparse to moderate late Tlamimilolpa, sparse to moderate Early Xolalpan, possible traces of Mazapan, and sparse Aztec components.

TC-73: 19, 685 m², slightly eroded. Artifactual remains indicated traces of Early Tlamimilolpa, moderate Late Tlamimilolpa, sparse Early Xolalpan, traces of Late Xolalpan, and sparse Aztec components.

TC-73:20, 690 m^2, slightly eroded. Artifactual remains indicated sparse Early Tlamimilolpa, moderate Late Tlamimilolpa, moderate Early Xolalpan, sparse Late Xolalpan, and sparse Aztec components.

(TC-73:1 through TC-73:20 are definitely identified on the survey maps, but TC-73:21 through TC-73:101 are identified only as concentrations within the site area. Some of these concentrations may be interpreted as platforms of other constructions and may not have been actual dwellings or pyramids at the site.)

TC-73:21, 1,440 m^2, slightly eroded. Artifactual remains indicated sparse Early Tlamimilolpa, moderate Late Tlamimilolpa, sparse Early Xolalpan, traces of Mazapan, and sparse Aztec components.

TC-73:22, 2,085 m^2, slightly eroded. Artifactual remains indicated traces of Early Tzacualli, traces of Early Tlamimilolpa, moderate Late Tlamimilolpa, moderate Early Xolalpan, traces of Late Xolalpan, and traces of Aztec components.

TC-73:23, 1,030 m^2, slightly eroded. Artifactual remains indicated sparse to moderate Early Tlamimilolpa, moderate to heavy Late Tlamimilolpa, traces of Early Xolalpan, and traces of Aztec components.

TC-73:24, 2,720 m^2, slightly eroded. Artifactual remains indicated moderate Early Tlamimilolpa, moderate Lake Tlamimilolpa, traces of Early Xolalpan, traces of Mazapan, and traces of Aztec components.

TC-73:25, 600+ m^2, heavily eroded and damaged. Artifactual remains indicated sparse Early Tlamimilolpa, heavy Late Tlamimilolpa, sparse Early Xolalpan, and traces of Aztec components.

TC-73:26, 1,365 m^2, slightly eroded. Artifactual remains indicated traces of Miccaotli, moderate Early Tlamimilolpa, moderate Late Tlamimilolpa, traces of Early Xolalpan, and traces of Aztec components.

TC-73:27, 985 m^2, slightly eroded. Artifactual remains indicated sparse Early Tlamimilolpa, heavy Late Tlamimilolpa, sparse Early Xolalpan, a sparse Aztec components.

TC-73:28, 920 m^2, slightly eroded. Artifactual remains indicated traces of Miccaotli, sparse to moderate Early Tlamimilolpa, moderate to heavy Late Tlamimilolpa, sparse Early Xolalpan, and sparse Aztec components.

TC-73:29, 3,445 m^2, slightly eroded. Artifactual remains indicated sparse Early Tlamimilolpa, moderate to heavy Late Tlamimilolpa, sparse Early Xolalpan, traces of Late Xolalpan, and traces of Aztec components.

TC-73:30, 1,500+ m^2, heavily eroded and damaged. Artifactual remains indicated traces of Early Tzacualli, traces of Early Tlamimilolpa, sparse to moderate Late Tlamimilolpa, moderate Early Xolalpan, traces of Late Xolalpan, and traces of Mazapan components.

TC-73:31, 1,080 m^2, slightly eroded. Units TC-73:9 and TC-73:11 are included within this larger concentration. Artifactual remains indicated traces of Early Tlamimilolpa, moderate Late Tlamimilolpa, moderate Early Xolalpan, sparse Late Xolalpan, and traces of Aztec components.

TC-73:32, 1,350 m^2, slightly eroded. Unit TC-73-10 is included in this concentration. Artifactual remains indicated traces of Early Tzacualli, sparse Early Tlamimilolpa, moderate to heavy Late Tlamimilolpa, sparse Early Xolalpan, and traces of Aztec components.

TC-73:33, 1,760 m^2, moderate to heavily eroded. Artifactual remains indicated traces of Miccaotli, traces of Early Tlamimilolpa, sparse to moderate Late Tlamimilolpa, moderate Early Xolalpan, sparse Late Xolalpan, traces of Mazapan, and traces of Aztec components.

TC-73:34, 6,320 m^2, heavily eroded. Artifactual remains indicated traces of Early Tlamimilolpa, sparse to moderate Late Tlamimilolpa, sparse Early Xolalpan, sparse Late Xolalpan, and traces of Aztec components.

TC-73:35, 4,545 m^2. Unit TC-73:19 is included within this concentration. Artifactual remains indicated traces of Early Tlamimilolpa, moderate Late Tlamimilolpa, moderate Early Xolalpan, sparse Late Xolalpan, traces of Mazapan, and traces of Aztec components.

TC-73:36, 2,200 m², slightly eroded. Unit TC-73:12 is included within this concentration. Artifactual remains indicated sparse Early Tlamimilolpa, moderate to heavy Late Tlamimilolpa, sparse Early Xolalpan, and traces of Late Xolalpan components. One unworked fragment of a <u>Chama echinata</u> or "Jewel Box" was found during surface survey. (See Appendices I, II and III.)

TC-73:37, 1,250 m². Artifactual remains indicated moderate Early Tlamimilolpa, moderate Late Tlamimilolpa, traces of Early Xolalpan, and traces of Aztec components.

TC-73:38, 3,700 m², heavily eroded and damaged. Artifactual remains indicated traces of Miccaotli, sparse to moderate Early Tlamimilolpa, moderate to heavy Late Tlamimilolpa, traces of Early Xolalpan, and sparse to moderate Aztec components.

TC-73:39, 695 m², slightly eroded. Artifactual remains indicated sparse Miccaotli, moderate Early Tlamimilolpa, moderate to heavy Late Tlamimilolpa, sparse Early Xolalpan, sparse to moderate Mazapan, and sparse Aztec components.

TC-73:40, 2,250 m², slightly eroded. Artifactual remains indicated traces of Miccaotli, sparse Early Tlamimilolpa, heavy Late Tlamimilolpa, traces of Early Xolalpan, traces of Mazapan, and traces of Aztec components.

TC-73:41, 4,050 m², heavily eroded. Unit TC-73:18 is included within this concentration. Artifactual remains indicated traces of Miccaotli, moderate to heavy Early Tlamimilolpa, moderate Late Tlamimilolpa, sparse Early Xolalpan, traces of Late Xolalpan, traces of Mazapan, and traces of Aztec components.

TC-73:42, 1,365 m², slightly eroded. Artifactual remains indicated traces of Miccaotli, sparse Early Tlamimilolpa, moderate Late Tlamimilolpa, sparse Early Xolalpan, traces of Late Xolalpan, traces of Mazapan, and traces of Aztec components.

TC-73:43, 3,850 m², slightly eroded. Artifactual remains indicated traces of Miccaotli, sparse to moderate Early Tlamimilolpa, heavy Late Tlamimilolpa, sparse Early Xolalpan, traces of late Xolalpan, sparse to moderate Mazapan, and sparse Aztec components.

TC-73:44, 3,690 m², heavily eroded. Artifactual remains indicated traces of Miccaotli, sparse to moderate Early Tlamimilolpa, moderate Late Tlamimilolpa, sparse Early Xolalpan, traces of Late Xolalpan, moderate Mazapan, and traces of Aztec components.

TC-73:45, 890 m^2, heavily eroded. Artifactual remains indicated traces of Miccaotli, sparse Early Tlamimilolpa, moderate to heavy Late Tlamimilolpa, moderate Early Xolalpan, traces of Late Xolalpan, moderate Mazapan, and traces of Aztec components.

TC-73:46, 1,190 m^2, slightly eroded. Artifactual remains indicated traces of Miccaotli, moderate to heavy Early Tlamimilolpa, heavy Late Tlamimilolpa, sparse Early Xolalpan, traces of Late Xolalpan, sparse to moderate Mazapan, and sparse Aztec components.

TC-73:47, 445 m^2, slightly eroded. Artifactual remains indicated traces of Miccaotli, sparse to moderate Early Tlamimilolpa, heavy Late Tlamimilolpa, sparse Early Xolalpan, traces of Late Xolalpan, sparse Mazapan, and traces of Aztec components.

TC-73:48, 610 m^2, slightly eroded. Artifactual remains indicated traces of Miccaotli, moderate Early Tlamimilolpa, heavy Late Tlamimilolpa, sparse Early Xolalpan, traces of Late Xolalpan, and traces of Aztec components.

TC-73:49, 5,556 m^2, slightly eroded. Units TC-73:16 and TC-73:17 are part of this concentration. Artifactual remains indicated traces of Miccaotli, sparse to moderate Early Tlamimilolpa, heavy Late Tlamimilolpa, sparse Early Xolalpan, traces of Late Xolalpan, moderate Mazapan, and sparse Aztec components.

TC-73:50, 1,600+ m^2, heavily eroded. Units TC-73:14 and TC-73:15 are part of this concentration. Artifactual remains indicated traces of Miccaotli, moderate to heavy Early Tlamimilolpa, heavy Late Tlamimilolpa, sparse Early Xolalpan, traces of Mazapan, and traces of Aztec components.

TC-73:51, 745 m^2, slightly eroded. Artifactual remains indicated traces of Miccaotli, moderate Early Tlamimilolpa, moderate Late Tlamimilolpa, traces of Early Xolalpan, possible traces of Late Xolalpan, traces of Mazapan, and sparse Aztec components.

TC-73:52, 1,590 m^2, slightly eroded. Artifactual remains indicated traces of Miccaotli, moderate Early Tlamimilolpa, heavy Late Tlamimilolpa, sparse Early

Xolalpan, traces of Late Xolalpan, traces of Mazapan, and sparse Aztec components.

TC-73:53, 560 m^2, heavily eroded Artifactual remains indicated traces of Miccaotli, moderate to heavy Early Tlamimilolpa, sparse Early Xolalpan, traces of Late Xolalpan, traces of Mazapan, and traces of Aztec components.

(TC-73:54 through TC-73:111, 11,250 m^2 total, heavily and damaged in most cases. These constitute 57 small house mounds on the TC-73 site.)

TC-73:54. Artifactual remains indicated traces of Miccaotli, moderate Early Tlamimilolpa, moderate to heavy Late Tlamimilolpa, sparse Early Xolalpan, traces of Late Xolalpan, traces of Mazapan, and sparse Aztec components.

TC-73:55. Artifactual remains indicated traces of Miccaotli, sparse Early Tlamimilolpa, moderate to heavy Late Tlamimilolpa, moderate Early Xolalpan, sparse Late Xolalpan, and traces of Aztec components.

TC-73:56. Artifactual remains indicated traces of Miccaotli, moderate Early Tlamimilolpa, heavy Late Tlamimilolpa, sparse Early Xolalpan, traces of Late Xolalpan, ans sparse Aztec components.

TC-73:57. Artifactual remains indicated sparse Early Tlamimilolpa, moderate to heavy Late Tlamimilolpa, moderate Early Xolalpan, traces of late Xolalpan, and traces of Aztec components.

TC-73:58. No sample was taken.

TC-73:59. Artifactual remains indicated traces of Miccaotli, sparse Early Tlamimilolpa, moderate to heavy Late Tlamimilolpa, sparse Early Xolalpan, traces of Late Xolalpan, and sparse Aztec components.

TC-73:60 through TC-73:101 were not sampled.

The site has a sparse to moderate but uneven distribution of lithic materials including obsidian blades, cores, projectile points, and scrapers. Ground stone tools, especially manos, metates, mortars, and pestles, were also noted. Other special lithic materials included a quartzite pebble pounder from TC-73:14.

Classification. TC-73 was interpreted by Marino (1965:141, 150, 164, 171-172, 186) in several ways. He considered the site to have Tlamimilolpa type apartment clusters as well as Xolalpan type apartment clusters within a District Center; in addition, he believed the site to be associated with a nearby canal irrigation system and also noted that it was also a Proto-Toltec living site. Marino (1975: 354-357), in considering the TT-134 component of this site, characterized it as a town and ceremonial center dating to the Mazapan or Late Toltec Phase. This judgment, however, is unclear especially in terms of the heavy Classic components at this site. Sanders (1965: 107, 118, 176) considered the site to be a small Classic town. This site is reclassified as a Provincial Center with a probable population between 787-1,570 during Xolalpan times. The site probably progressed from a Large Nucleated Village or Provincial Center during the Early Classic to a Provincial Center during the Middle Classic, and reverted to a smaller Provincial Center or Large Nucleated Village during the Late Classic. There is no evidence of Terminal Classic occupation at the site.

TC-91, TC-92, and TC-93

Background. Site TC-91, also known as San Marcos Ahuatepec Panteon, site TC-92, also known as San Marcos Ahuatepec el Centro, and site TC-93, also known as San Marcos Ahuatepec Oeste, were surveyed by Sanders in 1963. These sites, originally considered to be separate entities, have now been conjoined into a single site. The artifact collections from the Classic survey were processed by Kolb in 1964 and 1965 and consisted of a total of 11 samples. Collections made by the TA survey team were also studied by Kolb. Ceramic materials from the Classic survey were good to excellent for chronological purposes, while figurine data were good to excellent and the Classic field reports were generally of good quality. Published and unpublished materials on the site include Marino (1965) and Sanders (1965).

Natural Setting. Site TC-91/92/93 is located in the Upper Valley, East Piedmont, of the Teotihuacan Valley between 2,430-2,450 m in the Middle Piedmont Ecological Zone. Soils in the site area have a loamy texture and are light brown to dark brown in color, with depths ranging from 0-35+ cm. There is slight erosion in most of the site area, and tepetate is exposed in the northwestern and southeastern sections of the site. Moderate concentrations of rock and tezontle fragments are found in the area. Vegetation in the vicinity includes pirul, nopal, maguey, and various grasses.

Modern Utilization. Cultural features include no structures and no jagueys. The site area is used primarily for agricultural purposes including the cultivation of maguey and maize. Maguey bancals are also found in the vicinity. A modern highway is located to the north of the site.

Archaeological Remains. The total multicomponet site TC-91 covers 2.2 ha, while the Classic area covers 2.2 ha; at least one mound has been identified which has Classic occupation. The total multicomponent site TC-92 occupies 3.5 ha, while the Classic component occupies 3.3 ha; two mounds were identified of which both have Classic occupation. The total multicomponent site TC-93 occupies 0.2 ha, while the Classic component occupies 0.2 ha; one mound has been identified which has Classic occupation. The combined sites have five phases of the Preclassic, six phases of the Classic, and four phases of the Post Classic represented for a total of 15. Associated non-Classic sites include TF-143 and TA-211, while Classic sites TC-109 and TC-123 are located to the north and sites TC-94, TC-95, and TC-96 to the south. The following mounds were identified:

TC-91:1, 1,900 m², heavily eroded. Six samples were taken from this mound identified as TC-91:1a-f. In sum, the artifactual remains indicated traces of Cuanalan, traces of Patlachique, sparse Early Tzacualli, traces of Late Tzacualli, traces of Miccaotli, sparse to moderate Early Tlamimilolpa, moderate to heavy Late Tlamimilolpa, moderate Early Xolalpan, sparse to moderate Late Xolalpan, traces of Metepec, traces of Oxtotipac, traces of Xometla, sparse to moderate Mazapan, and sparse Aztec components. One unworked "Bittersweet Clam" fragment was recovered from TC-91:1b. (See Appendices I, II and III.)

TC-92:1, 1,850 m², heavily eroded. Four artifactual samples were taken from this site and were designated TC-92:A through TC-93:D. In sum, the artifactual remains indicated sparse Patlachique, heavy Early Tzacualli, traces of Late Tzacualli, traces of Miccaotli, possible traces of Early Tlamimilolpa, and traces of Late Tlamimilolpa components.

TC-93:1, 310 m², moderately to heavily eroded. Artifactual remains indicated moderate to heavy Early Tzacualli, traces of Late Tzacualli, traces of Miccaotli, and sparse Mazapan components.

The combined sites have an uneven and sparse distribution of lithic materials including obsidian blades, cores, and scrapers. Ground stone tools, especially manos and metates, were also noted. Other special lithic materials include a sedimentary shale hoe/chopper which illustrates pressure flaking and grinding techniques. This specimen was from collection TC-92:1c.

Classification. TC-91/92/93 were interpreted by Marino (1965:156) as a series of isolated multiroom houses of the Early Classic; Sanders (1965:119) regarded the site as a hamlet or village. The conjoined site TC-91/92/93 is reclassified as a Small Nucleated Village with a probable population between 270-540 during Xolalpan times. The site probably progressed from a Small Nucleated Village during the Early Classic to a Small Nucleated Village during the Middle Classic, and reverted to Hamlet status during the Late/Terminal Classic.

GLOSSARY OF ARCHITECTURAL TERMS

The following terms are used especially in Figures 8-15 and 19-23. Abbreviations used in these figures are in parentheses.

Alley (A): Unroofed access hallways leading from the Central Courtyard to apartments within rural residential units. These small "corridors" are normally 0.6 - 1.5 m in width.

Antesala (AS): A broad portico open to one side and having a roof supported by columns. The roof often has elaborate roof decorations. This architectural unit is found only in urban center palaces or associated with temples (such as the Palacio de Quetzalpapalotl).

Bench (B): A low, raised platform suitable for sitting or sleeping, normally located along the interior walls of rooms or along the walls of a Central Courtyard in residential units. Occasionally benches are found along side and/or rear walls of Porticos/Porches.

Central Courtyard (CC): Also called Principal Patio or Main Patio, this architectural unit is a large, rectilinear open-air area, usually surrounded by raised Porticos/Porches and room complexes in multi-family residences. The area is always paved with concrete/stucco and has a lime plastered surface. Courtyards are at levels below the Porticos/Porches and apatment Room(s), and served as a focal point of social, political, economic, and/or religious activities conducted in the residential unit.

Corridor (Co): An unroofed access hallway leading from the Central Courtyard to apartment or room complexes in urban residential units. These are normally 1.5 - 3.5 m in width and may have served as small "Streets."

Mural (m): Polychrome painted (mineral and organic pigments of black, red, red-orange, green, blue-green, yellow, and white) frescoes rendered on lime plastered walls of, especially, urban "elite" or "high status" residences and other structures. Murals always had geometric and naturalistic designs, and often had anthropomorphic, or zoomorphic, or human figure depictions, or a combination of these.

Patio (Pa): An unroofed, paved (usually) or unpaved internal open space within a house or apartment, or which served an apartment complex. Such an open-air area permitted light to enter and air to circulate, and it is assumed this area was used for social gatherings and were workplaces (similar to Roman and Etruscan atria). In urban residences, Patios often have central depressed areas called "lightwells" which serve as small water reservoirs; these normally have drainage systems to allow rainfall to be channelled away. Paving is similar to the Central Courtyard.

Platform: A large, raised structure found in the urban center, often having from two to four superimposed tiers of diminishing size from lower to upper tier, and accessed by a wide central stairway. The upper or terminal tier supported a temple or room complex (such as at the Palacio de Quetzalpapalotl).

Platform Altar (P): A raised platform, oten with talud and tablero (sloping talus supporting an entablature), normally located in a Central Courtyard. It is assumed that these "altars" served as foci of social and/or religious ceremonies practiced by the inhabitants of the surrounding residences or apartment units.

Portico/Porch (Po): A wide and shallow rectilinear space open to one side and having a roof supported by columns. This structure is situated in front of one or more Rooms, providing access to an apartment or house, and served as the intermediate unit between the Central Courtyard and apartment. Porticos are associated with urban residences, whereas the term Porch is used to designate such units in rural residences.

Room (R): A roofed architectural unit with four walls and at least one entryway. One or more Rooms composed an apartment entered through a Portico/Porch. Rooms are the most dominant architectural unit in both urban and rural residences.

Stairway (S): An architectural unit of from two to twelve steps providing access from a Central Courtyard to a Portico/Porch, Alley to Room, or Room to Patio. Most Stairways were balustraded and constructed of flat stones (lajas), concrete/stucco, and lime plaster. Wider, multi-stepped versions provided access to Antesala and Platforms in the urban center.

<u>Street</u> (St): An unroofed access "corridor" located between residential units in the urban center or between a residential unit and Platform. They are extremely variable in width, ranging from 2.5 to over 25.0 m. The Miccaotli ("Street of the Dead"), East/Eastern Avenue, and West/Western Avenue are the widest "Streets" in urban Teotihuacan. At least some streets were paved with concrete/stucco and had lime plaster surfaces.

GLOSSARY OF SPANISH AND NAHUATL TERMS

Agave americana: The maguey or American Century Plant, grown for its juice (agua miel) in the Meseta Central.

Agua miel: Literally "honey water," the sweet unfermented juice of the Agave americana which, when fermented, yields pulque, an alcoholic beverage.

Barranca: A deeply eroded gully, ravine, or gorge.

Barrio: A ward or district within or adjacent to an urban center; frequently a barrio was a unit of craft specialization or was an ethnic enclave.

Bodega: A storeroom or warehouse, a petalcalco (Nahuatl).

Caliche: A hard lens of white rock-like material, usually Calcium Carbonate ($CaCO_3$), deposited in the soils of the Meseta Central, especially in the Basin of Mexico.

Calpulli: A large patrilineal-based residential unit composed of multiple families having common sociopolitical, economic, and military functions (see Sanders and Price 1968:153-159 for further details).

Cascajo: Crushed volcanic tuff, ash, or cinder used in construction (see Tezontli).

Ejido: Communal land established after the 1910-1920 Mexican Revolution.

Granja: A small commercial farm, often specializing in a particular product such as poultry, berries, or fruit.

Huizache: The bush-like thorn tree, Acacia farnesiana.

Jaguey: An artificial water reservoir, often cut into bedrock or made by damming a stream or barranca.

Maguey: (See Agave americana.)

Meseta Central: The Mesoamerican Central Plateau, including the Basin of Mexico.

Opuntia: Cacti bearing the fruit called tuna (singular); Nopal are within the Opuntia biotic classification.

Pirul: Also called "Arbol de Peru," the *Shinus mole*, a common piedmont shade tree imported from Peru by the Spanish ca 1580.

Pulque: The fermented juice of the *Agave americana* or maguey, and an important alcoholic beverage from Prehispanic times to the present.

Talud and tablero: An architectural facade style characteristic of Classic Teotihuacan culture in Mesoamerica. It consists of a talus or sloping batter (talud) which supports an overhanging rectilinear entablature (tablero).

Tepetate: A fine-textured, indurated subsoil or basal "rock" of volcanic origin common in the Meseta Central.

Tezontli: Also spelled tezontle, a light volcanic tuff cut into blocks for construction or crushed to make cascajo.

Tierra caliente: Literally "hot land," the ecozone from sea level to ca 1000 m, with three subdivisions based on annual rainfall (see Sanders and Price 1968:104). Arid: below 800 mm of rainfall (northwestern Yucatan and the Isthmus of Tehuantepec); Subhumid: 800 - 1200 mm of rainfall (central Veracruz, coastal Oaxaca, central Chiapas Valley); and Humid: over 1200 mm of rainfall (the Peten, the Pacific coasts of Chiapas and Guatemala). Regions of tropical or subtropical flora and fauna.

Tierra fria: Literally "cold land," the ecozone from ca 2000 - 3000+ m, with three subdivisions based on annual rainfall (see Sanders and Price 1968:104). Arid: below 500 mm of rainfall (parts of Hidalgo and eastern Puebla); Subhumid: 500 - 1000 mm of rainfall (most of the Meseta Central); Humid: over 1000 mm of rainfall (higher slopes of the Highlands and escarpment areas).

Tierra templada: Literally "temperate land," the ecozone from ca 1000 - 2000 m, with three subdivisions based on annual rainfall (see Sanders and Price 1968:104). Arid: below 500 mm of rainfall (Meztitlan Valley and Tehuacan Valley); Subhumid: 500 - 1000 mm of rainfall (southern Puebla and Morelos, Valley of Oaxaca); Humid: over 1000 mm of rainfall (escarpments of the Pacific and Gulf Coastal Plains, Highland Guatemala).

Tlatel: a low mound, whether of artificial or natural origin (such as a geological drumlin).

BIBLIOGRAPHY

Abbott, R. Tucker

1954 *American Seashells*. New York: Van Nostrand.

Abbott, R. Tucker

1962 *Seashells of the World: A Guide to the Better-known Species*. New York: Western Publishing Co.

Abbott, R. Tucker

1968 *A Guide to Field Identification: Seashells of North America*. New York: Western Publishing Co.

Abbott, R. Tucker

1970 *How to Know the American Marine Shells*, rev. ed. New York: New American Library.

Abbott, R. Tucker

1975 *American Seashells*, 2nd ed. New York: Van Nostrand Reinhold.

Acosta, Jorge

1964 *El palacio del Quetzalpapalotl*. Mexico, D. F.: Instituto Nacional de Antropologia e Historia, *Memoria* 10.

Acosta Saignes, Miguel

1945 *Los pochteca: Ubicacion de los mercadores en la estructura social Tenochca*. Mexico, D. F.: Escuela Nacional de Antropologia e Historia, *Acta Antropologica* 1.

Adams, Richard E. W.

1971 *The Ceramics of Altar de Sacrificios*. Cambridge, Mass.: Harvard University, Peabody Museum of Archaeology and Ethnology, *Paper* 63(1).

Andrews, E. Wyllys, IV

1969 The Archaeological Use and Distribution of Mollusca in the Maya Lowlands. New Orleans: Tulane University, Middle American Research Institute, Publication 34.

Archivo General de Indias, Sevilla

1580 Archivo General de la Indias, Sevilla: Justicia. Sevilla. Leg. 208, No. 4, separate page, Fol. 3-7 (1530).

Archivo General de la Nacion, Mexico

1595 Archivo General de la Nacion, Mexico: Tierras. Mexico, D. F. Vol. 1520, Exp. 1, Fol. 6 (1595).

Armillas, Pedro

1950 Teotihuacan, Tula y los Toltecas: Las culturas post-archaicas y pre-Aztecas en el centro de Mexico: Excavaciones y estudios. Runa: Archivo para las Sciencias del Hombre 3:37-70.

Armytage, F.

1953 The Free Trade System in the British West Indies: A Study in Commercial Policy, 1766-1822. London: Longmans, Green.

Baker, Frank Collins

1932 Molluscan Shells of the Etowah Mounds. Andover, Mass.: Robert S. Peabody Foundation for Archaeology, Etowah Papers 5.

Ball, Joseph W. and Jack D. Eaton

1972 Marine Resources and the Prehistoric Lowland Maya: A Comment. American Anthropologist 74:772-776.

Barba de Piña Chan, Beatriz

1956 Tlapacoya: Un sitio preclasico de transicion. Mexico, D. F.: Escuela Nacional de Antropologia e Historia, Acta Antropologica Epoca 2, 1(1).

Barbour, Warren

    1976   The Figurines and Figurine Chronology of Ancient Teotihuacan. Rochester: University of Rochester, Ph.D. dissertation (University Microfilms 76-23,976).

Barlow, Robert H.

    1949   The Extent of the Empire of the Culhua Mexica. Berkeley and Los Angeles: University of California Press, Ibero-Americana 28.

Bell, Betty

    1971   Archaeology of Nayarit, Jalisco, and Colima. In Handbook of Middle American Indians, Vol. 11: Archaeology of Northern Mesoamerica, Part 2, Gordon F. Ekholm and Ignacio Bernal (eds.). Austin: University of Texas Press, 695-753.

Bell, Betty

    1974   Excavations at El Cerro Encantado, Jalico. In The Archaeology of West Mexico, Betty Bell (ed.). Ajijic, Jalisco: Sociedad de Estudios Avanzados el Occidente de Mexico, 147-167.

Bell, Betty (ed.)

    1974   The Archaeology of West Mexico. Ajijic, Jalisco: Sociedad de Estudios Avanzados del Occidente de Mexico.

Bell, Robert E.

    1947   Trade Materials at Spiro Mound as Indicated by Artifacts. American Antiquity 12:181-183.

Bennyhoff, James A.

    1966   Chronology and Periodization: Continuity and Change in the Teotihuacan Ceramic Tradition. In Onceava Mesa Redonda: Teotihuacan I. Mexico, D. F.: Sociedad Mexicana de Antropologia, 19-30.

Berdan, Frances Frei

    1978   Ports of Trade in Mesoamerica: A Reappraisal. In Mesoamerican Communication Routes and Cultural Contacts, Thomas A. Lee, Jr., and

Carlos Navarrete (eds.). Provo: New World Archaeological Foundation, Paper 40:187-198.

Berdan, Frances F.

1982 The Aztecs of Central Mexico: An Imperial Society. New York: Holt, Rinehart, and Winston.

Bernal, Ignacio

1951 Nuevos descubrimientos en Acapulco, Mexico. In The Civilizations of Ancient America: Selected Papers of the XXIXth International Congress of Americanists, New York, 1949, Sol Tax (ed.). Chicago: University of Chicago Press, 52-56.

Bernal, Ignacio (ed.)

1963 Teotihuacan: Descubrimientos, reconstrucciones. Mexico, D. F.: Instituto Nacional de Antropologia e Historia.

Beyer, Hermann

1933 Shell Ornament Sets from the Huasteca, Mexico. In Middle American Research Series, Studies in Middle America. New Orleans: Tulane University, Department of Middle American Research, Publication 5, Studies in Middle America 4:153-216.

Bilharz, Joy A.

1972 Bones, Bodies, and Diseases in Central Mexico. Paper presented at the Central States Anthropological Society Annual Meeting, Cleveland, Ohio.

Bittman, Bente and Thelma D. Sullivan

1978 The Pochteca. In Mesoamerican Communication Routes and Cultural Contacts, Thomas A. Lee, Jr., and Carlos Navarrete (eds.). Provo: New World Archaeological Foundation, Paper 40:211-218.

Boekelman, Henry, J.

1935 Ethno- and Archeo-Conchological Notes on Four Middle American Shells. Maya Research 2:257-277.

Boekelman, Henry J.

1936 Shell Trumpet from Arizona. American Antiquity 2:27-31.

Boekelman, Henry J.

1937 Two Probable Shell Trumpets from Ontario. American Antiquity 2:295-296.

Borhegyi, Stephan F. de

1961 Shark Teeth, Stingray Spines, and Shark Fishing in Ancient Mexico and Central America. Southwestern Journal of Anthropology 17:273-296.

Borhegyi, Stephan F. de

1965a Archaeological Synthesis of the Guatemalan Highlands. In Handbook of Middle American Indian, Vol. 2: Archaeology of Southern Mesoamerica, Part 1, Gordon R. Willey (ed.). Austin: University of Texas Press, 3-58.

Borhegyi, Stephan F. de

1965b Guatemalan Carbon-14 Dates. Katunob 5(2-3):33-34.

Borhegyi, Stephan F. de

1966a Shell Offerings and the Use of Shell Motifs at Lake Amatitlan, Guatemala, and Teotihuacan, Mexico. Actas y Memorias del XXXVI Congreso Internacional de Americanistas, Sevilla, 1966 1:355-371.

Borhegyi, Stephan F. de

1966b The Wind God's Breastplate. Expedition 8(4):13-15.

Borhegyi, Stephan F. de

1971 Pre-Columbian Contacts--The Dryland Approach: The Impact and Influence of Teotihuacan Culture on the Pre-Columbian Civilizations of Mesoamerica. In Man Across the Sea, Carroll L. Riley et al (eds.). Austin: University of Texas Press, 79-105.

Brand, Donald D.

1937 Southwestern Trade in Shell Products. American Antiquity 2:300-302.

Cardos, Amalia de Mendel

1959 El commercio de los Mayas antiguos. Mexico, D. F.: Escuela Nacional de Antropologia, Sociedad de Alumnos, Acta Antropologica Epoca 2, 2(1).

Carrasco, Pedro

1971 Social Organization of Ancient Mexico. In Handbook of Middle American Indians, Vol. 10: Archaeology of Northern Mesoamerica, Part 1, Gordon F. Ekholm and Ignacio Bernal (eds.). Austin: University of Texas Press, 349-375.

Caso, Alfonso

1965 Lapidary Work, Goldwork, and Copperwork from Oaxaca. In Handbook of Middle American Indians, Vol. 3: Archaeology of Southern Mesoamerica, Part 2, Gordon R. Willey (ed.) Austin: University of Texas Press, 896-930.

Castillo, Ignacio B. del

1922 Datos geograficos. In La poblacion del Valle de Teotihuacan, Tomo I, Vol. 2: La poblacion colonial, Manuel Gamio (ed.) Mexico, D. F.: Secretaria de Agricultura y Fomento, Direccion de Antropologia, 703-720.

Chadwick, Robert

1971 Archaeological Synthesis of Michocan and Adjacent Regions. In Handbook of Middle American Indians, Vol. 11: Archaeology of Northern Mesoamerica, Part 2, Gordon F. Ekholm and Ignacio Bernal (eds.). Austin: University of Texas Press, 657-693.

Chapman, Anne

1957 Port of Trade Enclaves in Aztec and Maya Civilization. In Trade and Market in the Early Empires, Karl Polanyi, Conrad Arensberg, and Harry Pearson (eds.) New York: Free Press, 114-153.

Chapman, Anne

    1959   Puertos de intercambio en Mesoamerica prehispanica. Mexico, D.F.: Instituto Nacional de Antropologia e Historia, Serie Historia 3.

Charlton, Thomas H.

    1972   The Significance of a Historic Event in the Archaeology of the Valley of Mexico. Paper presented at the Society of American Archaeology Annual Meeting, Bal Harbour, Florida.

Charlton, Thomas H.

    1978   Teotihuacan, Tepeapulco and Obsidian Exploitation. Science 200: 1227-1236.

Charlton, Thomas H.

    1979   Teotihuacan: Trade Routes of a Multi-tiered Economy. In Los processos de cambio: XV Mesa Redonda de la Sociedad Mexicana de Antropologia y Universidad de Guanajuato. Mexico, D. F.: Sociedad Mexicana de Antropologia, 285-292.

Charlton, Thomas H.

    1984   Production and Exchange: Variables in the Evolution of a Civilization. In Trade and Exchange in Early Mesoamerica, Kenneth G. Hirth (ed.). Albuquerque: University of New Mexico Press, 17-42.

Chavez O., Ernesto A.

    1969   Artifactual and non-artifactual Material of the Phyla Mollusca, Arthropoda, and Chordata from Chiapa de Corzo, Chiapas. In The Artifacts of Chiapa de Corzo, Chiapas, Mexico, Thomas A. Lee, Jr. Provo: New World Archaeological Foundation, Paper 26: 219-220.

Clark, J. C. (James Cooper Clark, trans, and ed.)

    1938   Cordex Mendoza: The Mexican Manuscript Known as the Collection of Mendoza and Preserved in the Bodleian Library, Oxford, 3 vols. London: Waterlow and Sons.

Clench, W. J.

    1947   The Genera Purpura and Thais in the Western Atlantic. Johnsonia 1(23):61-91.

Coan, Eugene V.

    1965   Kitchen Midden Mollusks of San Luis Gonzaga Bay. The Veliger: Publication of the California Malacozoological Society 7(4). (Reprinted in Katunob 6(4):42-45, 1965).

Cobean, Robert H.

    1978   The Pre-Aztec Ceramics of Tula, Hidalgo, Mexico. Cambridge, Mass.: Harvard University, Ph.D. dissertation.

Coe, Michael D.

    1965   Archaeological Synthesis of Southern Veracruz and Tabasco. In Handbook of Middle American Indians, Vol. 3: Archaeology of Southern Mesoamerica, Part 2, Gordon R. Willey (ed.). Austin: University of Texas Press, 679-715.

Coe, Michael D. and Kent V. Flannery

    1967   Early Cultures and Human Ecology in South Coastal Guatemala. Washington: Smithsonian Institution Press, Smithsonian Contributions to Anthropology 3.

Coe, William R.

    1959   Piedras Negras Archaeology: Artifacts, Caches, and Burials. Philadelphia: University Museum, University of Pennsylvania, Museum Monographs.

Coe, William R.

    1965a   Caches and Offeratory Practices of the Maya Lowlands. In Handbook of Middle American Indians, Vol. 2: Archaeology of Southern Mesoamerica, Part 1, Gordon R. Willey (ed.). Austin: University of Texas Press, 462-468.

Coe, William R.

    1965b   Artifacts of the Maya Lowlands. In Handbook of Middle American Indians, Vol. 3: Archaeology of Southern Mesoamerica, Part 2,

Gordon R. Willey (ed.). Austin: University of Texas Press, 594-602.

Coe, William

 1972 Cultural Contact between the Lowland Maya and Teotihuacan as Seen from Tikal, Peten, Guatemala. In <u>Onceava Mesa Redonda: Teotihuacan II</u>. Mexico, D. F.: Sociedad Mexicana de Antropologia, 257-271.

Cook de Leonard, Carmen

 1971a Ceramics of the Classic Period in Central Mexico. In <u>Handbook of Middle American Indians, Vol. 10: Archaeology of Northern Mesoamerica, Part 1</u>, Gordon F. Ekholm and Ignacio Bernal (eds.). Austin: University of Texas Press, 179-205.

Cook de Leonard, Carmen

 1971b Minor Arts of the Classic Period in Central Mexico. In <u>Handbook of Middle American Indians, Vol. 10: Archaeology of Northern Mesoamerica, Part 1</u>, Gordon F. Ekholm and Ignacio Bernal (eds.). Austin: University of Texas Press, 206-227.

Davidson, Judith R.

 1980 The Spondylus Shell: An Image of Ecological Regulation. Paper presented at the American Anthropological Association Annual Meeting, Washington, D. C.

Doran, Edwin, Jr.

 1958 The Cacaios Conch Trade. <u>Geographical Review</u> 48:389-401.

Drake, Robert J.

 1960 Nonmarine Molluscan Remains from an Archaeological Site at La Playa, Northern Sonora, Mexico. <u>Bulletin of the Southern California Academy of Science</u> 59:131-137.

Drennan, Robert D.

 1984a Long-distance Movement of Goods in the Mesoamerican Formative and Classic. <u>American Antiquity</u> 49:27-43.

Drennan, Robert D.

    1984b Long-Distance Transport Costs in Pre-Hispanic Mesoamerica. *American Anthropologist* 86:105-112.

Driver, Harold

    1969 *Indians of North America*, 2nd ed. Chicago: University of Chicago Press.

Drucker, Philip

    1965 *Cultures of the North Pacific Coast*. San Francisco: Chandler.

Earle, Timothy K. and Jonathon E. Ericson

    1977 Exchange Systems in Archaeological Perspective. In *Exchange Systems in Prehistory*, Timothy K. Earle and Jonathon E. Ericson (eds.). New York: Academic Press, 3-12.

Ekholm, Gordon F.

    1961 Some Collar-shaped Shell Pendants from Mesoamerica. In *Homenaje a Pablo Martinez del Rio en el XXV Aniversario de la Edicion de Los Origines Americanos*. Mexico, D.F.: n.p., 287-293.

Emerson, William K.

    1960 *Shell Middens of San Jose Island*. New York: American Museum of Natural History, *American Museum Novitates* 2013.

Enciso, Jorge

    1947 *Sellos del antiguo Mexico*. Mexico, D.F.: n.p.

Ester, Michael Ray

    1976 *The Spatial Allocation of Activities at Teotihuacan, Mexico*. Waltham, Mass.: Brandeis University, Ph.D. dissertation (University Microfilms 76-25,301).

Evans, J. G.

    1969 Land and Freshwater Mollusca in Archaeology: Chronological Aspects. *World Archaeology* 1(2):170-183.

Feldman, Lawrence H.

1968a   Maquixco Teotihuacan Archaeological Mollusks and Species List. University Park: Department of Anthropology, The Pennsylvania State University, Ms. on file.

Feldman, Lawrence H.

1968b   West Mexican Archaeological Molluscs. In Excavations at Tizapan el Alto, Jalisco, Clement W. Meighan and Leonard J. Foote. Los Angeles: University of California at Los Angeles, Latin American Center, Latin American Studies 11: 165-168.

Feldman, Lawrence H.

1969   Panamic Sites and Archaeological Mollusks of Lower California. The Veliger: Publication of the California Malacozoological Society 12: 165-168. (Reprinted in Katunob 7(3):1-4, 1969.)

Feldman, Lawrence H.

1972   Moluscos Mayas: Especies y origines. Estudios de Cultura Maya 8: 117-138.

Feldman, Lawrence H.

1974   Archaeomolluscan Species of Northwest Mesoamerica: Patterns of Natural and Cultural Distribution. In The Archaeology of West Mexico, Betty Bell (ed.). Ajijic, Jalisco: Sociedad de Estudios Avanzados del Occidente de Mexico, 225-239.

Feldman, Lawrence H.

1978   Comments. In Mesoamerican Communication Routes and Cultural Contacts, Thomas A. Lee, Jr. and Carlos Navarrete (eds.). Provo: New World Archaeological Foundation, Paper 40: 137-139.

Fernandez Navarrete, Martin et al

1842-   Coleccion de documentos ineditos para la
1895   historia de España, 112 vols. Madrid.

Fewkes, J. Walter

    1896   Pacific Coast Shells from Prehistoric Tusayan Pueblos. *American Anthropologist* 9:359-367.

Fisk, G.

    1967   *Marketing Systems: An Introductory Analysis.* New York: Harper and Row.

Flannery, Kent V.

    1968   The Olmec and the Valley of Oaxaca: A Model for Interregional Interaction in Formative Times. In *Dumbarton Oaks Conference on the Olmec*, Elizabeth P. Benson (ed.). Washington: Dumbarton Oaks Research Library and Collection, 179-217.

Flannery, Kent V.

    1976a   The Early Mesoamerican House. In *The Early Mesoamerican Village*, Kent V. Flannery (ed.). New York: Academic Press, 16-24.

Flannery, Kent V.

    1976b   Evolution of Complex Social Systems. In *The Early Mesoamerican Village*, Kent V. Flannery (ed.). New York: Academic Press, 162-173.

Fried, Morton H.

    1957   The Classification of Corporate Unilineal Descent Groups. *Journal of the Royal Anthropological Institute of Great Britain and Ireland* 87:1-29.

Furst, Peter

    1966   *Shaft Tombs, Shell Trumpets and Shamanism: A Culture Historical Approach to Problems in West Mexican Archaeology.* Los Angeles: University of California at Los Angeles, Ph.D. dissertation University Microfilms 66-11, 9007).

Gamio, Manuel

    1922a   Esculturas esquemorfas. In *La poblacion del Valle de Teotihuacan, Tomo I, Vol. 1: La poblacion prehispanica*, Manuel Gamio (ed.).

Mexico, D. F.: Secretaria de Agricultura y Fomento, Direccion de Antropologia, 196-200.

Gamio, Manuel

1922b Generalidades sobre la poblacion colonial. In La poblacion del Valle de Teotihuacan, Tomo I, Vol. 2: La poblacion colonial, Manuel Gamio (ed.). Mexico, D. F.: Secretaria de Agricultura y Fomento, Direccion de Antropologia, 367-393.

Gamio, Manuel

1922c Aspectos de la poblacion en el Siglo XIX. In La poblacion del Valle de Teotihuacan, Tomo I, Vol. 2: La poblacion colonial, Manuel Gamio (ed.). Mexico, D. F.: Secretaria de Agricultura y Fomento, Direccion de Antropologia, 735-771.

Garcia Cook, Angel

1981 The Historical Importance of Tlaxcala in the Cultural Development of the Central Highlands. In Supplement to the Handbook of Middle American Indians, Vol. 1: Archaeology, Jeremy A. Sabloff (ed.). Austin: University of Texas Press, 244-276.

Garcia Cook, Angel and Eila del Carmen Trejo

1977 Lo Teotihuacano en Tlaxcala. Communicaciones: Proyecto Puebla-Tlaxcala 14:54-70.

Garcia Icazabalecta, Joaquin (ed.)

1886- Nueva coleccion de documentos para la historia
1892 de Mexico, 3 vols. Mexico, D. F.

Gerhard, Peter

1946 Shellfish Dye in America. Actas y Memorias del XXXV Congresso Internacional de Americanistas, Mexico, 1962 3: 177-191.

Gibson, Charles

1964 The Aztecs under Spanish Rule. Stanford: Stanford University Press.

Gifford, E. W.

    1946   Archaeology in the Punta Peñasco Region, Sonora. American Antiquity 11:215-221.

Gifford, E. W.

    1947   California Shell Artifacts. Los Angeles and Berkeley: University of California, Anthropological Records 9(1).

Goggin, John M.

    1943   An Archaeological Survey of the Rio Tepalcatepec Basin, Michoacan. American Antiquity 9:44-58.

Greengo, Robert E.

    1954   Archaeological Marine Shells. In The Monagrillo Culture of Panama, Gordon R. Willey and Charles R. McGimsey. Cambridge, Mass.: Harvard University, Peabody Museum of American Archaeology and Ethnology, Papers 49(2):141-150.

Grove, David

    1968a   Chalcatzingo, Morelos, Mexico: A Reappraisal of the Olmec Rock Carvings. American Antiquity 33:486-491.

Grove, David

    1968b   The Morelos Pre-Classic and the Highand Olmec Problem: An Archaeological Study. Los Angeles: University of California at Los Angeles, Ph. D. dissertation (University Microfilms 68-11,869).

Grove, David

    1968c   The Pre-Classic Olmec in Central Mexico: Site Distribution and Inferences. In Dumbarton Oaks Conference on the Olmec, Elizabeth P. Benson (ed.). Washington: Dumbarton Oaks, Trustees for Harvard University, 179-185.

Harvey, Herbert R.

    1971   Ethnohistory of Guerrero. In Handbook of Middle American Indians, Vol. 10: Archaeology of Northern Mesoamerica, Part 1, Gordon F.

Ekholm and Ignacio Bernal (eds.). Austin: University of Texas Press, 603-618.

Herrera, Moises

    1922   Esculturas zoomorfas. In <u>La poblacion del Valle de Teotihuacan, Tomo I, Vol. 1: La poblacion prehispanica</u>, Manuel Gamio (ed.). Mexico, D. F.: Secretaria de Agricultura ya Fomento, Direccion de Antropologia, 187-190.

Hicks, Frederick and H. B. Nicholson

    1964   The Transition from Classic to Postclassic at Cerro Portezuelo, Valley of Mexico. <u>Actas y Memorias del XXXV Congreso Internacional de Americanistas, Mexico, D. F., 1962</u> 1:493-505.

Hirth, Kenneth G.

    1984a   Early Exchange in Mesoamerica: An Introduction In <u>Trade and Exchange in Early Mesoamerica</u>, Kenneth G. Hirth (ed.). Albuquerque: University of New Mexico Press, 1-15.

Hirth, Kenneth G.

    1984b   The Analysis of Prehistoric Economic Systems: A Look to the Future. In <u>Trade and Exchange in Early Mesoamerica</u>, Kenneth G. Hirth (ed.). Albuquerque: University of New Mexico Press, 281-302.

Hirth, Kenneth G. and Jorge Angulo Villaseñor

    1981   Early State Expansion in Central Mexico: Teotihuacan in Morelos. <u>Journal of Field Archaeology</u> 8:135-150.

Hodik, Barbara J.

    1974   <u>The Teotihuacan Craftsman as Dreamer, and Reflector: A Descriptive and Stylistic Analysis of Formative and Classic Period Figurines</u>. University Park: The Pennsylvania State University, Ph.D. dissertation (University Microfilms 75-19,769).

Hubbs, Carl and Gunnar I. Roden

    1964   Oceanography and Marine Life along the Pacific Coast. In <u>Handbook of Middle American Indians, Vol. 1: Natural Environment and Early</u>

Cultures, Robert C. West (ed.). Austin: University of Texas Press, 143-186.

Jackson, John Wilfrid

    1917   Shells as Evidence of the Migration of Early Culture. Manchester: University of Manchester, Publication 112, Ethnological Series 2.

Kaplan, Flora

    1961   A Shell from Mexico. El Mexico Antiguo 9:289-296.

Keeble, D. E.

    1967   Models of Economic Development. In Models in Geography, Richard J. Chorley and Peter Haggett (eds.). London: Methuen, 243-302.

Keen, A. Myra

    1960   Seashells of Tropical West America. Stanford: Stanford University Press.

Keen, A. Myra

    1971   Seashells of Tropical West America, 2nd ed. Stanford: Stanford University Press.

Kelley, J. Charles

    1971   Archaeology of the Northern Frontier: Zacatecas and Durango. In Handbook of Middle American Indians, Vol. 11: Archaeology of Northern Mesoamerica, Part 2, Gordon F. Ekholm and Ignacio Bernal (eds.). Austin: University of Texas Press, 768-801.

Kelly, Isabel T.

    1938   Excavations at Chametla, Sinaloa. Berkeley: University of California, Ibero-Americana 14.

Kelly, Isabel T.

    1945a   The Archaeology of the Autlan-Tuxcacuesco Area of Jalisco, I: The Autlan Zone. Berkeley and Los Angeles: University of California, Ibero-Americana 26.

Kelly, Isabel T.

    1945b   Excavations at Culiacan, Sinaloa. Berkeley and Los Angeles: University of California, Ibero-Americana 25.

Kelly, Isabel T.

    1947   Excavations at Apatzingan, Michoacan. New York: Wenner-Gren Foundation for Anthropological Research, Viking Fund Publications in Anthropology 7.

Kidder, Alfred V.

    1947   The Artifacts of Uaxactun, Guatemala. Washington: Carnegie Institution of Washington, Publication 576.

Kidder, Alfred V., Jesse D. Jennings, and Edwin M. Shook

    1946   Excavations at Kaminaljuyu, Guatemala. Washington: Carnegie Institution of Washington, Publication 561.

Kirchhoff, Paul

    1943   Mesoamerica: Sus limites geograficos, composicion etnica y caracteres culturales. Acta Americana 1(1):92-107.

Kolb, Charles C.

    1962   Field Notes: Santa Maria Maquixco el Bajo Excavations (TC-8). University Park: Department of Sociology and Anthropology, The Pennsylvania State University, Ms. on file.

Kolb, Charles C.

    1963   Teotihuacan Valley Project: Classic Site Surveys (TC):1963. University Park: Department of Sociology and Anthropology, The Pennsylvania State University, Site Reports on file.

Kolb, Charles C.

    1964   La Ventilla A-B Palacio-Monticulo Excavations. Rochester: Department of Anthropology, University of Rochester (Teotihuacan Mapping Project), Ms. on file.

Kolb, Charles C.

1965a   A Tentative Ceramics Classification for the Teotihuacan Valley (Patlachique through Aztec V Phases). University Park: Department of Anthropology, The Pennsylvania State University, Ms. on file.

Kolb, Charles C.

1965b   Ceramic Analysis of the Classic Sites in the Valley of Teotihuacan. University Park: Department of Anthropology. The Pennsylvania State University, Ms. on file

Kolb, Charles C.

1970   Classic Teotihuacan Figurines from he Teotihuacan Valley, Mexico. Paper presented at the Society for American Archaeology Annual Meeting, Mexico, D.F.

Kolb, Charles, C.

1972a   Lowland Maya and Oaxacan Influences in the Teotihuacan Valley. Paper presented at the Society for American Archaeology Annual Meeting, Bal Harbour, Florida.

Kolb, Charles C.

1972b   Field Notes: Teotihuacan Valley, Tlaxcala-Puebla Basin, 1972 Field Season. University Park: Department of Anthropology, The Pennsylvania State University, Ms. on file.

Kolb. Charles C.

1973a   The Old Shell Game: A Mesoamerican Trade Network. Paper presented at the Society for American Archaeology Annual Meeting, San Francisco, California.

Kolb, Charles C.

1973b   Classic Teotihuacan Figurines. University Park: Department of Anthropology, The Pennsylvania State University, Ms. on file.

Kolb, Charles C.

1973c   Thin Orange Pottery at Teotihuacan. In Miscellaneous Papers in Anthropology, William

T. Sanders (ed.). University Park: Department of Anthropology, The Pennsylvania State University, Occasional Papers in Anthropology 8:309-377.

Kolb, Charles C.

    1977   Technological Investigations of Mesoamerican 'Thin Orange' Ceramics. Current Anthropology 18:534-536.

Kolb, Charles C.

    1979a   Classic Teotihuacan Settlement Patterns in the Teotihuacan Valley, Mexico, 2 vols. University Park: The Pennsylvania State University, Ph.D. dissertation (University Microfilms 80-10,074).

Kolb, Charles C.

    1979b   The Classic Teotihuacan Period Chronology: Some Reflections. Katunob 11(2):1-51.

Kolb, Charles C.

    1982   Ceramic Technology and Problems and Prospects of Provenience in Specific Ceramics from Mexico and Afghanistan. In Archaeological Ceramics, Jacqueline S. Olin and Alan D. Franklin (eds.). Washington: Smithsonian Institution Press, 193-208.

Kolb, Charles C.

    1984a   Technological and Cultural Aspects of Teotihuacan Period 'Thin Orange' Ware. In Pots and Potters: Current Approaches to Ceramic Archaeology, Prudence M. Rice (ed.). Los Angeles: Unversity of California at Los Angeles Institute of Archaeology Press, Monograph 24:209-226.

Kolb, Charles C.

    1984b   Classic Teotihuacan "Granular Wares" -- Ceramic Technological and Cultural Interpretations. University Park: Department of Anthropology, The Pennsylvania State University, Ms. on file.

Kolb, Charles C.

1985 Demographic Estimates in Archaeology: Contributions from Ethnoarchaeology on Mesoamerican Peasants. Current Anthropology 26(5):581-599.

Kolb, Charles C.

1986 Commercial Aspects of Classic Teotihuacan Period 'Thin Orange' Ware. In Research in Economic Anthropology, Barry L. Isaac (ed.). Greenwich, Conn.: JAI Press, 155-205.

Kolb, Charles C.

1987a The Cultural Ecology of Classic Teotihuacan Copoid Ceramics. In A Pot for All Reasons: Ceramic Ecology Revisited, Charles C. Kolb and Louana M. Lackey (eds.). Philadelphia: Temple University, Laboratory of Anthropology, Occasional Paper 2, in press.

Kolb, Charles C.

1987b Classic Teotihuacan Period "Copoid Wares" -- Ceramic Technological and Cultural Interpretations. Ceramica de Cultura Maya et al. (whole number), in press.

Kolb, Charles C. and Joy A. Bilharz

1972 Paleopathology in Rural Teotihuacan. Paper presented at the Society for American Archaeology Annual Meeting, Bal Harbour, Florida.

Krickeberg, Walter

1956 Altmexikanishe Kulturen. Berlin: Safari-Verlag.

Krickeberg, Walter

1964 Las antiguas culturas mexicanas, 2nd ed. Mexico, D. F.: Fondo de Cultura Economica.

Krotser, Paul and Evelyn Rattray

1980 Manufactura y distribucion de tres grupos ceramicos de Teotihuacan. Anales de Antropologia 17:95-104.

Lange, Frederick W.

    1971   Marine Resources: A Viable Subsistence Alternative for the Prehistoric Lowland Maya. American Anthropologist 73:619-639.

Lee, Thomas A., Jr. and Carlos Navarrete (eds.)

    1978   Mesoamerican Communication Routes and Cultural Contacts. Provo: New World Archaeological Foundation, Paper 40.

Leechman, Douglas

    1949   Suggested Use of Clam Shells. American Antiquity 15(1):56.

Linne, Sigvald

    1934   Archaeological Researches at Teotihuacan, Mexico. Stockholm: Ethnographical Museum of Sweden, n.s. Publication 1.

Linne, Sigvald

    1942   Mexican Highland Cultures: Archaeological Researches at Teotihuacan, Calpulalpan, and Chalchicomula in 1934/35. Stockholm: Ethnographical Museum of Sweden, n.s. Publication 7.

Lister, Robert H.

    1947   Archaeology of the Middle Rio Balsas Basin, Mexico. American Antiquity 13:67-78.

Lister, Robert H.

    1949   Excavations at Cojumatlan, Michoacan. Albuquerque: University of New Mexico Press.

Litvak King, Jamie

    1978   Central Mexico as a Part of the General Mesoamerican Communications System. In Mesoamerican Communication Routes and Cultural Contacts, Thomas A. Lee, Jr. and Carlos Navarrete (eds.). Provo: New World Archaeological Foundation, Paper 40:115-122.

Lorenzo, Jose L.

    1955   Los concheros de la Costa de Chiapas. *Anales del Instituto Nacional de Antropologia e Historia* 7: 41-50.

Lowe, Gareth W. and J. Alden Mason

    1965   Archaeological Survey of the Chiapas Coast, Highlands, and Upper Grijalva Basin. In *Handbook of Middle American Indians, Vol. 2: Archaeology of Southern Mesoamerica, Part 1*, Gordon R. Willey (ed.). Austin: University of Texas Press, 195-236.

MacNeish, Richard S.

    1971   Archaeological Synthesis of the Sierra. In *Handbook of Middle American Indians, Vol. 11: Archaeology of Northern Mesoamerica, Part 2*, Gordon F. Ekholm and Ignacio Bernal (eds.). Austin: University of Texas Press, 573-581.

MacNeish, Richard S., Antoinette Nelken-Terner, and Irmgard W. Johnson

    1967   *The Prehistory of the Tehuacan Valley, Vol. 2: Non-ceramic Artifacts*. Austin: University of Texas Press.

Mahler, Joy

    1965   Garments and Textiles of the Maya Lowlands. In *Handbook of Middle American Indians, Vol. 3: Archaeology of Southern Mesoamerica, Part 2*, Gordon R. Willey (ed.). Austin: University of Texas Press, 581-593.

Malouf, Carling

    1940   Prehistoric Exchange in the Northern Periphery of the Southwest. *American Antiquity* 6: 115-122.

Margain, Carlos R.

    1971   Pre-Columbian Architecture of Central Mexico. In *Handbook of Middle American Indians, Vol. 10: Archaeology of Northern Mesoamerica, Part 1*, Gordon F. Ekholm and Ignacio Bernal (eds.). Austin: University of Texas Press, 45-91.

Marino, Joseph D., Jr.

    1965   *Settlement Types of the Formative and Classic Periods in the Teotihuacan Valley, Mexico.* University Park: The Pennsylvania State University, M. A. thesis.

Marino, Joseph D., Jr.

    1975   *Toltec Settlement Patterns in the Teotihuacan Valley, Mexico.* University Park: The Pennsylvania State University, Ph.D. dissertation (University Microfilms 76-17,192).

Marquina, Ignacio

    1922   Arquitectura contemporanea. In *La poblacion del Valle de Teotihuacan, Tomo II: La poblacion contemporanea*, Manuel Gamio (ed.). Mexico, D. F.: Secretaria de Agricultura y Fomento, Direccion de Antropologia, 575-594.

Marquina, Ignacio

    1951   *Arquitectura Prehispanica.* Mexico, D. F.: Instituto Nacional de Antropologia e Historia, *Memoria* 1.

Martens, Edward von

    1899   Purpur-Farberei in Central America. *Berliner Gesellschaft fur Anthropologie Verhandlungen, Zeitschrift fur Ethnologie* 1898: 482-486.

Matteson, Max R.

    1959   Snails in Archeological Sites. *American Anthropologist* 61: 1094-1096.

Mayer-Oakes, William J.

    1959   A Stratigraphic Excavation at El Risco, Mexico. *Proceedings of the American Philosophical Society* 103: 332-373.

Meighan, Clement W.

    1971   Archaeology of Sinaloa. In *Handbook of Middle American Indians*, ol. 11: *Archaeology of Northern Mesoamerica, Part 2*, Gordon F. Ekholm and Ignacio Bernal (eds.). Austin: University of Texas Press, 754-767.

Meighan, Clement W. and Leonard J. Foote

    1968   Excavations at Tizapan el Alto, Jalisco. Los Angeles: University of California at Los Angeles, Latin American Center, Latin American Studies 11.

Miles, S. W.

    1965   Summary of Preconquest Ethnology of the Guatemala-Chiapas Highlands and Pacific Slopes. In Handbook of Middle American Indians, Vol. 2: Archaeology of Southern Mesoamerica, Part 1, Gordon R. Willey (ed.). Austin: University of Texas Press 276-287.

Miller, Arthur G.

    1973   The Mural Painting of Teotihuacan. Washington: Dumbarton Oaks/Trustees for Harvard University.

Miller, Arthur G.

    1978   A Brief Outline of the Artistic Evidence for Classic Period Cultural Contact between Maya Lowlands and Central Mexican Highlands. In Middle Classic Mesoamerica: A. D. 400-700, Esther Pasztory (ed.). New York: Columbia University Press, 63-70.

Millon, Rene

    1964   The Teotihuacan Mapping Project. American Antiquity 29:345-352.

Millon, Rene

    1966a   Cronologia y periodificacion: Datos estratigraficos sobre periodos ceramicos y sus relaciones con la pintura mural. In Onceava Mesa Redonda: Teotihuacan I. Mexico, D. F.: Sociedad Mexicana de Antropologia, 1-18.

Millon, Rene

    1966b   Extension y poblacion en la ciudad de Teotihuacan en sus diferentes periodos. In Onceava Mesa Redonda: Teotihuacan I. Mexico, D. F.: Sociedad Mexicana de Antropologia, 57-78.

Millon, Rene

    1967a   Teotihuacan. <u>Scientific American</u> 216(6):38-63.

Millon, Rene

    1967b   Urna de Monte Alban III-A encontrada en Teotihuacan. <u>Boletin del Instituto Nacional de Antropologia</u> 29:42-44.

Millon, Rene

    1970   Teotihuacan: Completion of Map of Giant Ancient City in the Valley of Mexico. <u>Science</u> 170:1077-1082.

Millon, Rene

    1973   <u>Urbanization at Teotihuacan, Vol. 1, Parts 1-2: The Teotihuacan Map.</u> (Text and Maps.) Austin: University of Texas Press.

Millon, Rene

    1976   Social Relations in Ancient Teotihuacan. In <u>The Valley of Mexico</u>, Eric R. Wolf (ed.). Albuquerque: University of New Mexico Press, 205-248.

Millon, Rene

    1981   Teotihuacan: City, State, and Civilization. In <u>Supplement to the Handbook of Middle American Indians, Vol. 1: Archaeology</u>, Jeremy A. Sabloff (ed.). Austin: University of Texas Press, 198-243.

Moholy-Nagy, Hattula

    1963   Shells and Other Marine Material from Tikal. <u>Estudios de Cultura Maya</u> 3:65-83.

Monzon, Arturo

    1949   <u>El capulli en la organizacion social de los Tenochca.</u> Mexico, D. F.: Universidad Nacional Autonoma de Mexico, Instituto de Historia, Serie 1, <u>Publicacion</u> 14.

Moore, Clarence B.

1921 Notes on Shell Implements from Florida. *American Anthropologist* 23:12-18.

Morris, Percy A.

1966 *A Field Guide to Shells of the Pacific Coast and Hawaii*, 2nd ed. Boston: Houghton Mifflin.

Mountjoy, Joseph B.

1978 Prehispanic Cultural Contact on the South-central Coast of Nayarit, Mexico. In *Mesoamerican Communication Routes and Cultural Contacts*, Thomas A. Lee, Jr. and Carlos Navarrete (eds.). Provo: New World Archaeological Foundation, *Paper* 40:127-139.

Müller, E. Florencia Jacobs

1966a Secuencia ceramica de Teotihuacan. In *Onceava Mesa Redonda: Teotihuacan I*. Mexico, D. F.: Sociedad Mexicana de Antropologia, 31-44.

Müller, E. Florencia Jacobs

1966b Instrumental y armas. In *Onceava Mesa Redonda: Teotihuacan I*. Mexico, D. F.: Sociedad Mexicana de Antropologia, 225-238.

Müller, E. Florencia Jacobs *et al*

1963 *Mesa redonda de la ceramica de Teotihuacan: 13 sesions*. University Park: Department of Anthropology, The Pennsylvania State University, Ms. on file.

Murdock, George Peter

1949 *Social Structure*. New York: Macmillan.

Nicholson, Henry B.

1971 Major Sculpture in Pre-Hispanic Central Mexico. In *Handbook of Middle American Indians, Vol. 10: Archaeology of Northern Mesoamerica, Part 1*, Gordon F. Ekholm and Ignacio Bernal (eds.). Austin: University of Texas Press, 92-134.

Noguera, Eduardo

    1944   Exploraciones en Jiquilpan. <u>Anales del Museo Michoacano</u> 3: 37-52.

Noguera, Eduardo

    1955   Extraordinario hallazgo en Teotihuacan. <u>El Mexico Antiguo</u> 8: 43-56.

Noguera, Eduardo

    1965   <u>La ceramica arqueologica de Mesoamerica</u>. Mexico, D. F.: Universidad Nacional Autonoma de Mexico, Instituto de Investigaciones Historicas.

Noguera, Eduardo

    1971   Minor Arts in the Central Valleys. In <u>Handbook of Middle American Indians, Vol. 10: Archaeology of Northern Mesoamerica, Part 1</u>, Gordon F. Ekholm and Ignacio Bernal (eds.). Austin: University of Texas Press, 258-269.

Nuttall, Zelia

    1909   A Curious Survival in Mexico of the Use of the <u>Purpura</u> Shellfish for Dying. In <u>Putnam Anniversary Volume: Essays Presented to Frederick Ward Putnam ... by His Friends and Associates</u>. New York: n.p., 368-384.

Nuttall, Zelia

    1926   <u>Official Reports on the Towns of Tequizistlan, Tepechpan, Acolman, and San Juan Teotihuacan Sent by Francisco de Casteñada to His Majesty Philip II, and the Council of the Indies, in 1580</u>. Cambridge, Mass.: Harvard University, Peabody Museum of American Archaeology and Ethnology, <u>Papers</u> 11(2):42-84.

Oviedo y Valdes, G. F. (J. Amador de los Rios, ed.)

    1851-   <u>Historia general y natural de las Indias,</u>
    1856   <u>islas y tierra-firme del mar oceano</u> 4 tomos. Madrid: n.p.

Parker, R. H.

    1964   Zoogeography and Ecology: Macro-Invertebrates, Gulf of California and Continental Slope off

Mexico. Videnskabelige Meddelelser Fiza Dansk Naturhistorisk Forening 126:2-178.

Parmelee, Paul W. and Walter E. Klippel

1974 Freshwater Mussels as a Prehistoric Food Resource. American Antiquity 39:421-434.

Parsons, Jeffrey R.

1966 The Aztec Ceramic Sequence in the Teotihuacan Valley, Mexico, 2 vols. Ann Arbor: University of Michigan, Ph.D. dissertation.

Parsons, Jeffrey R.

1971 Prehispanic Settlement Patterns in the Texcoco Region, Mexico. Ann Arbor: University of Michigan, Museum of Anthropology, Memoir 3.

Parsons, Jeffrey R., Elizabeth Brumfiel, Mary Parsons, and David J. Wilson

1982 Prehistoric Settlement Patterns in the Southern Valley of Mexico: The Chalco-Xochimilco Region. Ann Arbor: University of Michigan, Museum of Anthropology, Memoir 14.

Parsons, Lee A.

1978 The Peripheral Coastal Lowlands and the Middle Classic Period. In Middle Classic Mesoamerica: A. D. 400-700, Esther Pasztory (ed.). New York: Columbia University Press, 25-34.

Parsons, Lee A. and Barbara J. Price

1971 Mesoamerican Trade and Its Role in the Emergence of Civilization. Contributions of the University of California Archaeological Research Facility 11:169-195.

Paso y Troncoso, Francisco del (ed.)

1905-1948 Papeles de Nueva España, Vol. 6: Relaciones geograficas de la diocesis de Mexico, 1579-1582 (1580). Madrid y Mexico, D. F.

Pasztory, Esther

1978 Historical Synthesis of the Middle Classi Period. In Middle Classic Mesoamerica: A. D.

400-700, Esther Pasztory (ed.). New York: Columbia University Press, 3-22.

Paulsen, Allison

    1974   The Thorny Oyster and the Voice of God: Spondylus and Strombus in Andean Prehistory. American Antiquity 39:597-607.

Peñafiel, Antonio

    1890   Libro de tributos: Monumentos del artes mexicano antiguo, 2 tomos. Berlin: A. Ascher.

Piña Chan, Roman

    1963   Excavaciones en el Rancho 'La Ventilla.' In Teotihuacan: Descubrimientos, reconstrucciones, Ignacio Bernal (ed.). Mexico, D. F.: Instituto Nacional de Antropologia e Historia, 50-52.

Piña Chan, Roman

    1971   Preclassic or Formative Pottery and Minor Arts of the Valley of Mexico. In Handbook of Middle American Indians, Vol. 10: Archaeology of Northern Mesoamerica, Part 1, Gordon F. Ekholm and Ignacio Bernal (eds.). Austin: University of Texas Press, 157-178.

Pires-Ferreira, Jane Wheeler

    1978   Shell Exchange Networks in Formative Mesoamerica. In Cultural Continuity in Mesoamerica, David L. Browman (ed.). The Hague: Mouton Publishers, 79-100.

Pollock, Harry E. D.

    1965   Architecture of the Maya Lowlands. In Handbook of Middle American Indians, Vol. 2: Archaeology of Southern Mesoamerica, Part 1, Gordon R. Willey (ed.). Austin: University of Texas Press, 378-440.

Porter, Muriel Noe

    1956   Excavations at Chupicuaro, Guanajuato, Mexico. Philadelphia: American Philosophical Society, Transactions 46(5).

Proskouriakoff, Tatiana

    1962   The Artifacts of Mayapan. In *Mayapan, Yucatan, Mexico*, H. E. D. Pollock, Ralph L. Roys, T. Proskouriakoff, and A. Ledyard Smith. Washington: Carnegie Institution of Washington, *Publication* 619:321-438.

Rands, Robert L

    1965   Jades of the Maya Lowlands. In *Handbook of Middle American Indians, Vol. 3: Archaeology of Southern Mesoamerica, Part 2*, Gordon R. Willey (ed.). Austin: University of Texas Press, 561-580.

Rands, Robert L.

    1969   *Mayan Ecology and Trade: 1967-1968*. Carbondale: Southern Illinois University, University Museum, *Mesoamerican Studies* 2.

Rands, Robert L. and Robert E. Smith

    1965   Pottery of the Guatemalan Highlands. In *Handbook of Middle American Indians, Vol. 2: Archaeology of Southern Mesoamerica, Part 1*, Gordon R. Willey (ed.). Austin: University of Texas Press, 95-145.

Rathje, William L., David A. Gregory, and Frederick M. Wiseman

    1978   Trade Models and Archaeological Problems: Classic Maya Examples. In *Mesoamerican Communication Routes and Cultural Contacts*, Thomas A. Lees, Jr. and Carlos Navarrete (eds.). Provo: New World Archaeological Foundation, *Paper* 40:147-175.

Rattray, Evelyn Childs

    1973   *The Teotihuacan Ceramic Chronology: Early Tzacualli to Early Tlamimilolpa Phases*. Columbia: University of Missouri, Ph.D. dissertation (University Microfilms 74-18,619).

Real Academia de la Historia

    1898   *Documentos ineditos relaciones de Yucatan*, segunda serie, *Tomo* 11. Madrid: Real Academia de la Historia.

Richards, Horace G. and H. J. Boekelman

    1937   Shells from Maya Excavations in British Honduras. American Antiquity 3:166-169.

Ricketson, O. G. and E. B. Ricketson

    1937   Uaxactun, Guatemala: Group E-1926-31. Washington: Carnegie Institution of Washington, Publication 477.

Roys, Ralph L.

    1943   The Indian Background of Colonial Yucatan. Washington: Carnegie Institution of Washington, Publication 548.

Roys, Ralph L.

    1965   Lowland Maya Native Society at Spanish Contact. In Handbook of Middle American Indians, ol. 3: Archaeology of Southern Mesoamerica, Part 2, Gordon R. Willey (ed.). Austin: University of Texas Press, 659-678.

Rubin de la Borbolla, D. F.

    1947   Teotihuacan: Ofrenda de los templos de Quetzalcoatl. Anales del Instituto Nacional de Antropologia e Historia 2:61-72.

Sabloff, Jeremy A. and David A. Freidel

    1975   A Model of a Pre-Columbian Trading Center. In Ancient Civilization and Trade, Jeremy A. Sabloff and C. C. Lamberg-Karlovsky (eds.). Albuquerque: University of New Mexico Press, 369-408.

Sabloff, Jeremy A. and William L. Rathje

    1975a   A Study of Changing Pre-Columbian Commercial Systems: The 1972-1973 Seasons at Cozumel, Mexico. Cambridge, Mass.: Harvard University, Peabody Museum of Archaeology and Ethnology, Monograph 3.

Sabloff, Jeremy A. and Wiliam L. Rathje

    1975b   The Rise of a Maya Merchant Class. Scientific American 233(4):72-82.

Safer, Jane Fearer and Frances McLaughlin Gill

    1982   <u>Spirals from the Sea: An Anthropological Look at Shells</u>. New York: Clarkson N. Potter in association with the American Museum of Natural History.

Sahagun, Fray Bernardino de (Arthur J. O. Anderson and Charles E. Dibble, trans.)

    1950-   <u>Florentine Codex: General History of the</u>
    1969   <u>Things of New, Spain</u>, 12 books. Santa Fe: University of Utah and School of American Research.

         1951 Book 2: <u>The Ceremonies</u>

         1952 Book 3: <u>The Origin of the Gods</u>

         1954 Book 8: <u>Kings and Lords</u>

         1959 Book 9: <u>The Merchants</u>

         1961 Book 10: <u>The People</u>

         1963 Book 11: <u>Earthly Things</u>

Sanders, William T.

    1965   <u>The Cultural Ecology of the Teotihuacan Valley</u>. University Park: Department of Sociology and Anthropology, The Pennsylvania State University.

Sanders, William T.

    1966   Life in a Classic Village. In <u>Onceava Mesa Redonda: Teotihuacan I</u>. Mexico, D. F.: Sociedad Mexicana de Antropologia, 123-148.

Sanders, William T.

    1978   Ethnographic Analogy and the Teotihuacan Horizon Style. In <u>Middle Classic Mesoamerica: A. D. 400-700</u>, Esther Pasztory (ed.). New York: Columbia University Press, 35-44.

Sanders, William T. and Barbara J. Price

    1968   <u>Mesoamerica: The Evolution of a Civilization</u>. New York: Random House.

Sanders, William T. et al

    1970   The Teotihuacan Valley Project Final Report, Vol. 1: The Natural Environment, Contemporary Occupation, and 16th Century Population. University Park: Department of Anthropology, The Pennsylvania State University, Occasional Papers in Anthropology 3.

Sanders, William T., Jeffrey R. Parsons, and Robert S. Santley

    1979   The Basin of Mexico: The Cultural Ecology of a Civilization. New York: Academic Press.

Sanders, William T. and David Webster

    1978   Unilinealism, Multilinealism, and the Evolution of Complex Societies. In Social Archaeology: Beyond Subsistence and Dating, Charles R. Redman (ed.). New York: Academic Press, 249-302.

Sanders, William T., Deborah Nichols, Rebecca Storey and Randolph Widmer

    1982   A Reconstruction of a Classic Period Landscape in the Teotihuacan Valley. University Park: The Pennsylvania State University, Department of Anthropology, Final Report to the National Science Foundation (BNS 8005754).

Santley, Robert S.

    1977   Intra-site Settlement Patterns at Loma Torremote and their Relationship to Formative Prehistory in the Cuautitlan Region, State of Mexico. University Park: The Pennsylvania State University, Ph.D. dissertation (University Microfilms 78-08419).

Santley, Robert S.

    1984   Obsidian Exchange, Economic Stratification, and the Evolution of Complex Society in the Basin of Mexico. In Trade and Exchange in Early Mesoamerica, Kenneth G. Hirth (ed.). Albuquerque: University of New Mexico Press, 43-86.

Santley, Robert S., Ponciano Ortiz Ceballos, Thomas W. Kellion, Philip J. Arnold and Janet M. Kerley

    1984   Final Field Report of Matacapan Archaeological Project: The 1982 Season. Albuquerque: University of New Mexico, Latin American institute, Research Paper 15.

Santley, Robert S., Ponciano Ortiz Ceballos, Philip J. Arnold, Ronald A. Kneebone, Michael P. Smyth and Janet M. Kerley

    1985   Final Field Report of Matacapan Project: The 1983 Season. Albuquerque: University of New Mexico, Department of Anthropology, Ms.

Schenck, W. Egbert and E. W. Gifford

    1952   Archaeological Sites on Opposite Shores of the Gulf of California. American Antiquity 17:265.

Sejourne, Laurette

    1959   Un palacio en la ciudad de los dioses: Exploraciones en Teotihuacan, 1955-1958. Mexico, D. F.: Instituto Nacional de Antropologia e Historia.

Sejourne, Laurette

    1966a   Arqueologia de Teotihuacan: La ceramica. Mexico, D. F.: Fondo de Cultura Economica.

Sejourne, Laurette

    1966b   Arquitectura y pintura en Teotihuacan. Mexico, D. F.: Siglo Ventiuno Editores, S. A.

Sejourne, Laurette

    1966c   El lenguaje de las formas en Teotihuacan. Mexico, D. F.: Litoarte.

Sejourne, Laurette

    1969   Teotihuacan: Metropole de l'Amerique. Paris: Francois Maspero.

Seler, Eduard

    1915   Die Teotihuacan-Kultur des Hochlands von Mexico. Gesammelte Abhandlungen zur

Americanischen Sprach- und Alterthumskunde 5: 405-585.

Seler, Eduard

    1963   Commentarios al Codice Borgia, 2 tomos. Mexico, D. F.: Fondo de Cultura Economica.

Shackleton, Nicholas and Colin Renfrew

    1970   Neolithic Trade Routes Re-aligned by Oxygen Isotope Analyses. Nature 228: 1062-1065.

Shook, Edwin M.

    1965   Archeological Survey of the Pacific Coast of Guatemala. In Handbook of Middle American Indians, Vol. 2: Archaeology of Southern Mesoamerica, Part 1, Gordon R. Willey (ed.). Austin: University of Texas Press, 180-194.

Siliceo Pauer, Paul

    1925   Representaciones prehispanicas de dientes humanos hechas en concha. Anales del Museo Nacional de Mexico Epoca 4, 3: 210-222.

Silva-Galdames, Osvaldo

    1971   Trade and the Concepts of Nuclear and Marginal Culture Areas in Mesoamerica. Ceramica de Cultura Maya 7(Supplement): 1-74.

Smith, A. Ledyard and Alfred V. Kidder, with T. D. Stewart

    1951   Excavations at Nebaj, Guatemala. Washington: Carnegie Institution of Washington, Publication 594.

Spence, Michael W.

    1971   Skeletal Morphology and Social Organization in Teotihuacan, Mexico. Carbondale: Southern Illinois University, Ph.D. dissertation (University Microfilms 72-10,302).

Spence, Michael W.

    1976   Human Skeletal Material from the Oaxaca Barrio in Teotihuacan, Mexico. In Archaeological Frontiers: Papers on New World High Cultures in Honor of J. Charles Kelley, Robert B. Pickering (ed.). Carbondale: Southern

Illinois University, <u>University Museum Studies</u> 4:129-148.

Spence, Michael W.

1984 Craft Production and Polity in Early Teotihuacan. In <u>Trade and Exchange in Early Mesoamerica</u>, Kenneth G. Hirth (ed.). Albuquerque: University of New Mexico Press, 87-114.

Starbuck, David R.

1975 <u>Man-Animal Relationships in Pre-Columbian Central Mexico</u>. New Haven: Yale University, Ph. D. dissertation (University Microfilms 75-24,604).

Starbuck, David R.

1977 Animal Utilization and Urban Adaptations in the City of Teotihuacan, Mexico. <u>Western Canadian Journal of Anthropology</u> 7:151-162.

Starr, Frederick

1897 A Shell Gorget from Mexico. <u>The Antiquarian</u> 1(3):57-61.

Storey, Rebecca and Randolph Widmer

1982 Excavations at Tlajinga 33. In <u>A Reconstruction of a Classic Period Landscape in the Teotihuacan Valley</u>. University Park: The Pennsylvania State University, Department of Anthropology, Final Report to the National Science Foundation (BNS 8005754), 21-97.

Stresser-Pean, Guy

1971 Ancient Sources on the Huasteca. In <u>Handbook of Middle American Indians, Vol. 11: Archaeology of Northern Mesoamerica, Part 2</u>, Gordon F. Ekholm and Ignacio Bernal (eds.). Austin: University of Texas Press, 582-618.

Stromsvik, Gustav

1942 Substela Caches and Stela Foundations at Copan and Quirigua. In <u>Contributions to American Anthropology and History</u>. Washington: Carnegie Institution of Washington, Publication 528, 7(37):63-96.

Thomas, R. S.

    1956   Free Ports and Foreign Trade Zones. Ithaca: Cornell University, Cornell Maritime Press.

Thompson, John E. S.

    1939   Excavations at San Jose, British Honduras. Washington: Carnegie Institution of Washington, Publication 506.

Thompson, J. Eric S.

    1964   Trade Relations between the Maya Highlands and Lowlands. Estudios de Cultura Maya 4: 13-50.

Thompson, J. Eric S.

    1965   Archaeological Synthesis of the Southern Maya Lowlands. In Handbook of Middle American Indians, Vol. 2: Archaeology of Southern Mesoamerica, Part 1, Gordon R. Willey (ed.). Austin: University of Texas Press, 331-359.

Tolstoy, Paul B.

    1958   Surface Survey of the Northern Valley of Mexico: The Classic and Postclassic Periods. Philadelphia: American Philosophical Society, Transactions 48(5).

Tortellot, Gair and Jeremy a Sabloff

    1972   Exchange Systems among the Ancient Maya. American Antiquity 37: 126-135.

Tower, Donald H.

    1945   The Use of Marine Mollusca and Their Value in Reconstructing Prehistoric Trade Routes in the American Southwest. Cambridge, Mass.: Excavators Club, Papers 2(3).

Tozzer, Alfred M.

    1921   Excavation of a Site at Santiago Ahuizotla, D. F., Mexico. Washington: Smithsonian Institution, Bureau of American Ethnology, Bulletin 74.

Tozzer, Alfred M.

    1957   Chichen Itza and Its Cenote of Sacrifice: A Comparative Study of Contemporaneous Maya and

Toltec. Cambridge, Mass.: Harvard University, Peabody Museum of American Archaeology and Ethnology, Memoirs 11-12.

Vaillant, George C.

1930 Excavations at Zacatenco. New York: American Museum of Natural History, Anthropological Papers 32(1).

Vaillant, George C.

1931 Excavations at Ticoman. New York: American Museum of Natural History, Anthropological Papers 32(1).

Vaillant, George C.

1934 Excavations at El Corral, Santiago Ahuizotla (Azcapotzalco), Mexico: Field Notes, December 3-21, 1934. New York: Department of Anthropology, American Museum of Natural History, Ms. on file.

Vaillant, George C.

1935a Excavations at El Arbolillo. New York: American Museum of Natural History, Anthropological Papers 35(2).

Vaillant, George C.

1935b Excavations at Santa Maria Chiconautla, Mexico: Field Notes, February-March, 1935. New York: Department of Anthropology, American Museum of Natural History, Ms. on file.

Vaillant, Suzannah B. and George C. Vaillant

1934 Excavations at Gualupita. New York: American Museum of Natural History, Anthropological Papers 35(1).

Vance, James E., Jr.

1971 The Mechant's World: The Geography of Wholesaling. Englewood Cliffs, New Jersey: Prentice-Hall.

Villagra Caleti, Agustin

1951 Las pinturas de Atetelco en Teotihuacan. Cuadernos Americanos 10: 153-162.

Villagra Caleti, Agustin

1952 Teotihuacan: Sus pinturas murales. Anales del Instituto Nacional de Antropologia e Historia 5: 67-74.

Villagra Caleti, Agustin

1954 Trabajos realizados en Teotihuacan, 1952. Anales del Instituto Nacional de Antropologia e Historia 6: 69-78.

Villagra Caleti, Agustin

1956- Las pinturas de Atetelco, Teotihuacan.
1957 Boletin del Instituto Nacional de Antropologia e Historia 4: 1-3.

Villagra Caleti, Agustin

1971 Mural Painting in Central Mexico. In Handbook of Middle American Indians, Vol. 10: Archaeology of Northern Mesoamerica, Part 1, Gordon F. Ekholm and Ignacio Bernal (eds.). Austin: University of Texas Press, 135-156.

Vokes, Elizabeth H.

1963 A Possible Hindu Influence at Teotihuacan. American Antiquity 29: 94-95.

von Winning, Hasso

1947 A Symbol for Dripping Water in the Teotihuacan Culture. El Mexico Antiguo 6: 333-341.

von Winning, Hasso

1949 Shell Designs on Teotihuacan Pottery. El Mexico Antiguo 7: 126-153.

Warmke, Germaine L. and R. Tucker Abbott

1961 Caribbean Seashells. Narberth, Penna.: Livingston Publishing Co.

Weigand, Phil C.

1974 The Ahualulco Site and the Shaft-Tomb Complex of the Etzatlan Area. In The Archaeology of West Mexico, Betty Bell (ed.). Ajijic, Jalisco: Sociedad de Estudios Avanzados del Occidente de Mexico, 120-131.

Weitlaner Johnson, Irmgard

    1971   Basketry and Textiles. In *Handbook of Middle American Indians, Vol. 10: Archaeology of Northern Mesoamerica, Part 1*, Gordon F. Ekholm and Ignacio Bernal (eds.). Austin: University of Texas Press, 297-321.

Willey, Gordon R. and William R. Bullard, Jr.

    1965   Prehistoric Settlement Patterns in the Maya Lowlands. In *Handbook of Middle American Indians, Vol. 2: Archaeology of Southern Mesoamerica, Part 1*, Gordon R. Willey (ed.). Austin: University of Texas Press, 360-377.

Willey, Gordon R. and William R. Bullard, Jr., John B. Glass, and James C. Gifford, with others

    1965   *Prehistoric Maya Settlements in the Belize Valley*. Cambridge, Mass.: Harvard University, Peabody Museum of Archaeology and Ethnology, *Paper* 54.

Winter, Marcus C.

    1973   *Tierras Largas: A Formative Community in the Valley of Oaxaca*. Tucson: University of Arizona, Ph.D. dissertation.

Woodbury, Richard B. and Aubrey S. Trik

    1954   *The Ruins of Zaculeu, Guatemala*, 2 vols. Richmond, Virginia: William Byrd Press for the United Fruit Company.

Woodward, Arthur

    1936   A Shell Bracelet Manufactory. *American Antiquity* 2:117-125.

Zeiller, Warren

    1974   *Tropical Marine Invertebrates of Southern Florida and the Bahama Islands*. New York: Wiley-Interscience.

Zeitlan, Robert N.

    1978   Long-Distance Exchange and the Growth of a Regional Center on the Southern Isthmus of Tehuantepec. In *Prehistoric Coastal Adaptations: The Economy and Ecology of Maritime Middle America*, Barbara L. Stark and

Barbara Voohies (eds.). New York: Academic Press, 183-210.

PERSONAL COMMUNICATIONS

Adams, Richard E. W.

    1972   May 5, 1972.

Culbert, T. Patrick

    1972   May 5, 1972.

Ekholm, Gordon F.

    1973   June 6-24, 1970.

Feldman, Lawrence H.

    1973   May 5-6, 1973.

Feinberg, Harold S.

    1971   March 26 and July 18-20, 1971.

    1972   July 21, 1972.

    1973   April 13, 1973.

Gifford, James

    1971   April 19, 1971.

    1972   March 20, 1972.

Rands, Robert

    1972   May 6, 1972.

Santley, Robert S.

    1984   September 3, 1984.

Sonin, Robert

    1972   May 5-6, 1972.

Weigand, Phil C.

    1972   May 5 and October 2, 1972.

ABOUT THE AUTHOR

Charles C. Kolb is currently Director of Research and Grants at Mercyhurst College in Erie, Pennsylvania, U.S.A. He did all of his undergraduate and graduate work at The Pennsylvania State University, University Park, Pennsylvania, having taken his B.A. in Latin American History, with minors in Archaeology and Art History. In 1979, he received his doctorate in Latin American Archaeology and Anthropology, having written his dissertation on the Classic Teotihuacan Period settlement patterns in the Teotihuacan Valley, Basin of Mexico (Kolb 1979a). As an undergraduate and a young graduate student, he was one of the five original members of William T. Sanders' "Teotihuacan Valley Project," the first settlement pattern study conducted in North America, and later was an associate of Rene F. Millon's "Teotihuacan Mapping Project" (The University of Rochester). In 1962, Maurice A. Mook and Kolb excavated the TC-8:3 site, and in 1963-1964 Kolb surveyed the Classic Period Teotihuacan Valley sites from the Delta through Middle Valley. From 1962-1969, he was responsible for processing and classifying all ceramic and other artifacts from 134 Classic sites in the Valley. He did additional field work in the Basins of Mexico, Puebla, and Tlaxcala in 1970 and 1972.

Kolb was co-director and ceramic specialist with Louis Dupree's "Prehistoric Research in Afghanistan Project" for the American Museum of Natural History (1965-1967), and has authored a number of papers and articles on the pottery from four sites he excavated near Aq Kupruk, Afghanistan. In addition, he conducted extensive site surveys and excavated 22 archaeological sites in Pennsylvania (1967-date), and did ecological and archaeological surveys in Iron Age sites in Uganda, East Africa (1970).

He has taught at The Pennsylvania State University Main Campus, Bryn Mawr College, and The Behrend College of The Pennsylvania State University (Erie, Pennsylvania), and now occasionally teaches courses at Mercyhurst College. The author of over sixty articles and book or monograph chapters, and 22 book and film reviews, he has also presented more than eighty papers at professional archaeology meetings. Kolb is a Fellow of both the Royal Anthropological Institute and the American Anthropological Association, and is a member of the American Association for the Advancement of Science, American Ethnological Society, American Society for Ethnohistory, Archaeological Institute of America, Association for Field Archaeology, Society for American

Archaeology, Society for Archaeological Sciences, Society for Historical Archaeology, and 24 state, regional, or local archaeological or historical societies or associations. He is a founding member of the Paleopathology Association and an annually certified member of the Society of Professional Archaeologists since that organization began in 1977. In addition, Kolb was elected to Alpha Kappa Delta (Sociology), Phi Kappa Phi (Scholastic), Pi Gamma Mu (Social Sciences), the Society of the Sigma Xi (Scientific), and the New York Academy of Sciences. He served as an elected member of The Pennsylvania Humanities Council an affiliate of the National Endowment for the Humanities (1979-1983 and 1986-1990), was the Council's Secretary-Treasurer (1980-1982), and was and is a permanent member of the Fundraising and Finance committees (1981-1983 and 1986-1990).

His particular research interests include ceramic ecology (technology and typology), ethnoarchaeology, paleodemography, ethnohistory (archival and cartographic), settlement pattern studies, Cultural Resource Management, and pedagogy. Kolb's research areas include Mesoamerica, Central Asia and the Asian Subcontinent, and the Lower Great Lakes Basin (especially the States of Pennsylvania, New York, and Ohio). He also served as a consultant for the United States Department of Agriculture Forest Service and the United States Army Corps of Engineers/Department of the Interior.